R

S

FE PRESS

New York

Pocket Book Guides for Quant Interviews

1. 150 Most Frequently Asked Questions on Quant Interviews, by Dan Stefanica, Radoš Radoičić, and Tai-Ho Wang. FE Press, 2013

2. 150 Most Frequently Asked Questions on Quant Interviews (Chinese-English Edition), by Dan Stefanica, Radoš Radoičić, and Tai-Ho Wang with Jun Hua and Yu Gan. FE Press, 2017

3. 50 Challenging Brainteasers from Quant Interviews, by Radoš Radoičić, Ivan Matić, Dan Stefanica. FE Press, 2017

4. Stochastic Calculus & Probability Quant Interview Questions, by Ivan Matić, Radoš Radoičić, and Dan Stefanica. FE Press, 2017

Other Titles from FE Press

1. A Primer for the Mathematics of Financial Engineering, Second Edition, by Dan Stefanica. FE Press, 2011

2. Solutions Manual – A Primer for the Mathematics of Financial Engineering, Second Edition, by Dan Stefanica. FE Press, 2011

3. Numerical Linear Algebra Primer for Financial Engineering, by Dan Stefanica. FE Press, 2014

4. Solutions Manual – Numerical Linear Algebra Primer for Financial Engineering, by Dan Stefanica. FE Press, 2014

150 MOST FREQUENTLY ASKED QUESTIONS ON QUANT INTERVIEWS

150道最常见的金融工程面试题

DAN STEFANICA

RADOŠ RADOIČIĆ

TAI-HO WANG

Baruch College

City University of New York

with

JUN HUA and YU GAN

FE Press

New York

FE PRESS
New York

www.fepress.org

This edition first published 2017

Printed in the United States of America

ISBN-13 978-0-9797576-8-6
ISBN-10 0-9797576-8-1

Chinese translation by Jun Hua and Yu Gan

Cover art and design by Max Rumyantsev

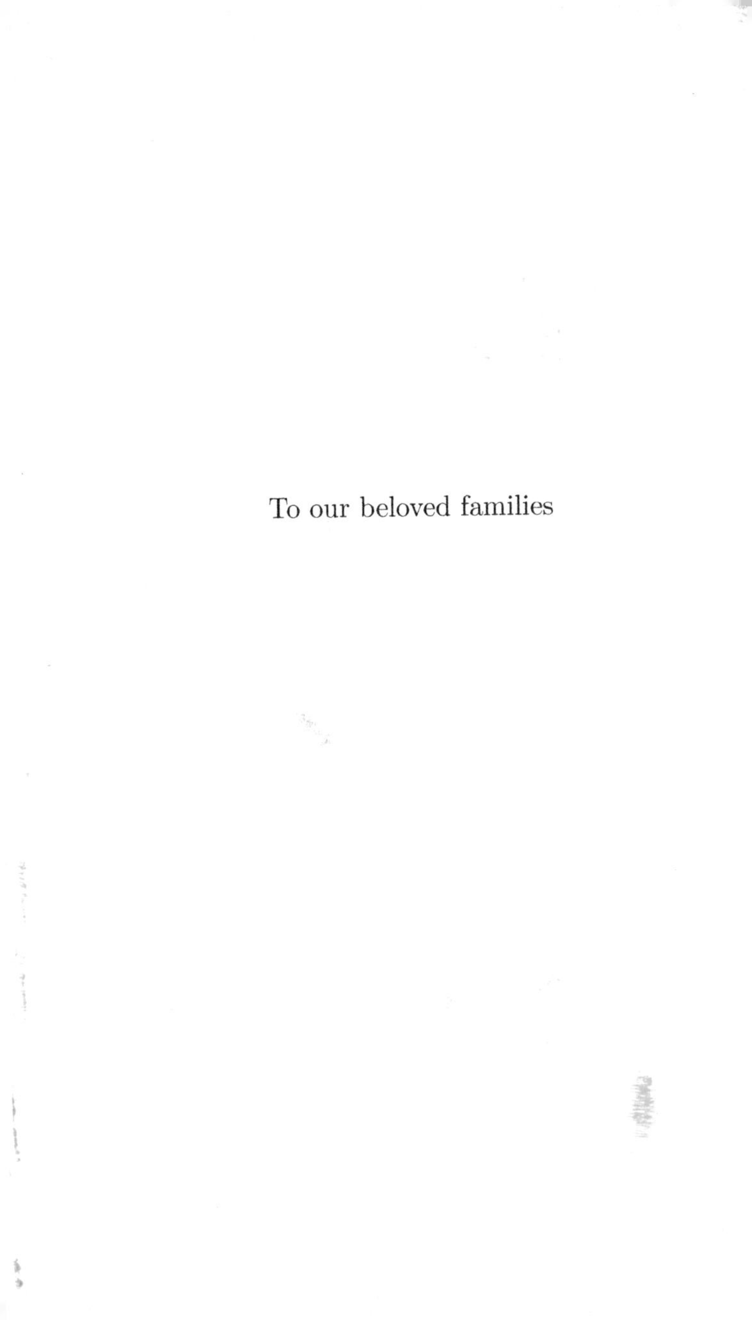

To our beloved families

目录

中英双语版前言

近年来在北美，来到巴鲁克学院（作者在此任教超过十年）以及其他金融工程硕士项目学习的中国学生人数迎来了大幅增长。

为了应对中国学生的增加，以及尽可能地帮助以中文为主的学生和其他读者准备求职面试，我们决定推出我们在2013年出版的畅销书《150道最常见的金融工程面试题》的中英双语版。

为了让读者方便地参照英文原文以及每题答案的英文表述，双语版包括了英文版中每一题的英文原题和相应的中文翻译。

与此同时，我们希望这本书能帮助读者在北美面试量化分析师岗位时脱颖而出。这将是对我们最好的肯定。

这本《150道最常见的金融工程面试题》是金工面试口袋指南丛书的第一本书。在此之后我们还将推出《概率和随机微积分面试题》和《金工脑经急转弯》。

纽约, 2017

Preface to the Chinese/English Bilingual Edition

The number of students of Chinese descent studying in financial engineering programs in United States, including in the Baruch College Financial Engineering Masters Program where the authors have been faculty members for over ten years, has grown tremendously in recent years. To acknowledge this increase and to potentially help these students (and other Chinese readers) in preparing for job interviews, we decided to generate a bilingual Chinese/English version of the best-selling "150 Most Frequently Asked Questions on Quant Interviews" book we published in 2013.

Every questions from the original book is included in this edition, with both its English version and its Chinese translation. We made the choice of having this bilingual delivery of the answers to further help the readers by making referencing the corresponding English terms easy, and including effective English deliveries of the answers to the questions.

We would like to think that the book will facilitate

success for the readers of this book on interviews for quant positions in United States, which would make our efforts worthwhile.

The book ““150 Most Frequently Asked Questions on Quant Interviews” is the first book in the *Pocket Book Guides for Quant Interviews* Series, to be followed by books on advanced probability and stochastic calculus questions and on challenging brainteasers asked in quant interviews.

New York, 2017

英文版前言

近年来数值方法与编程技能在金融领域中，尤其是交易和风险管理方面的应用迅速增长，并且不久前的金融危机和大数据时代的到来也加快了这个趋势。

量化交易相关的职位需要丰富和扎实的知识储备。而这些知识囊括了金融，计算机编程(以C++为主)，数学的多个领域(概率论，统计微分，数值分析，线性代数，高等微积分)。与此同时，脑筋急转弯也常常被用来考察面试者是否有独特的思维能力。

这本书中收录了能够帮助你进入最后一轮面试必不可少的150道题目。这些题目涵盖了上述所有核心知识点，它们不仅在过去而且现在也常常被量化交易相关职位的面试官反复引用。

同时，读者也可以找到这些题目的详细解答。这些解答是非常全面和直达要点的，是真实面试中的满分答案。

本书包括的主题有:

- 基础数学,微积分,微分方程.
- 协方差矩阵与相关系数矩阵. 线性代数.
- 金融产品: 期权,债券,掉期,远期,期货.
- C++. 数据结构.
- 蒙特卡洛模拟. 数值方法.
- 概率论. 随机分析.
- 脑筋急转弯.

本书作者是巴鲁克学院金融工程项目多位拥有近20年教学经验的资深教师。他们辅导过的学生大部分都在面试中取得优异的成绩并且取得了非常优秀的量化交易职位。正因为这样，本书也有幸得到了这些优秀校友们的帮助。这本书也是向巴鲁克学院金融工程项目的致敬。

这本书是*量化面试指导手册* 系列丛书的第一本。我们即将推出高等概率论和随机微分相关的面试题集和高难脑筋急转弯面试题集。

二〇一七年，纽约

Preface to the English Edition

The use of quantitative methods and programming skills in all areas of finance, from trading to risk management has grown tremendously in recent years, and accelerated through the financial crisis and with the advent of the big data era.

A core body of knowledge is required for successfully interviewing for a quant type position. The challenge lies in the fact that this knowledge encompasses finance, programming (in particular C++ programming), and several areas of mathematics (probability and stochastic calculus, numerical methods, linear algebra, and advanced calculus). Moreover, brainteasers are often asked to probe the ingenuity of candidates.

This book contains over 150 questions covering this core body of knowledge, without which it is not possible to advance to a final interview round. These questions are not only frequently, but also currently, asked on interviews for quantitative positions, and cover a vast spectrum, from C++ and data structures, to finance, brainteasers, and stochastic calculus.

The answers to all of these questions are included in the book. These answers are written in the same very practical vein that was used to select the questions: they are complete, but straight to the point – as they would be given in an interview.

Topics:

- Mathematics, calculus, differential equations.
- Covariance and correlation matrices. Linear algebra.
- Financial instruments: options, bonds, swaps, forwards, futures.
- C++, algorithms, data structures.
- Monte Carlo simulations. Numerical methods.
- Probability. Stochastic calculus.
- Brainteasers.

The authors are faculty members of the Baruch College Financial Engineering Masters Program, and have over 20 years of experience educating students who were very successful interviewing for quantitative positions. As such, the authors had the privilege to interact with generations of exceptional students, whose contributions as alumni to the continued success of our students has been tremendous. This book is a tribute to our special Baruch MFE community.

This is the first book in the *Pocket Book Guides for Quant Interviews* Series, to be followed by books on advanced probability and stochastic calculus questions and on challenging brainteasers asked in quant interviews.

New York, 2013

致谢

作为巴鲁克学院金融工程硕士项目的教授，我们一直对于能够为众多杰出学生、校友们的早期职业生涯成功提供帮助而深感荣幸。同时，围绕着巴鲁克金融工程硕士项目所成长起来的茁壮团体也是教育从业者们真心想要看到的。而这些对我们来说，也是一种鼓舞。如果没有我们的巴鲁克金融工程硕士项目团体的帮助，这本书将不会有机会与大家见面。因此，我们很感激每一个参与本书编辑的人。在此尤其要感谢我们的校友华俊（巴鲁克14届）和甘雨（巴鲁克14届），此书双语版的面世离不开他们日以夜继的努力。尤其需要提到的是，他们一丝不苟以及追求卓越的品质不仅帮助他们在职业生涯中取得了巨大的成功，同时他们也将这种态度应用于翻译原版书中的每一个细节，以及最佳的用词选择中。一如既往地，此书的封面是由我们的校友，同时也是朋友的Max Rumyantsev所设计。

此外，我们也要感谢许多学生主动为双语版提供的校对帮助，他们是：Huiyou (Francis) Chen，Zhihao (Alex) Chen， Mingyang (Corbin) Guan，Zijun (Alfred) Huang，Weihao (Vincent) Li，Sheng (Matthew) Lin，Yuchen (Derek) Qi， Zilun Shen，Xinlu Xiao，Gongshun (Gordon) Yin，Mengqi (Kiki) Zhang，Mengyu (Kristin) Zhang，以及ShengQuan Zhou。同时，我们还要感谢在原版书中提供校对帮助的校友们，他们是首先是最重要的Alejandro Canete，然后是需要再次感谢的Yu Gan和Jun Hua，以及Alireza Kashef，Yi Bill Lu，Svetlana Rafailova，Fubo Shi，Yujia

Helen Sun，Jun Charlotte Wang，Peng Wu，Yanzhu Wendy Wu，Yongyi Ivan Ye， He Hillary Zhao，和 Wenyi Zhou。如果缺少了这样一个大家族中任意一员所提供的支持与理解，这本书都将失去与大家见面的机会。因此，我们对他们需要抱有寸草春晖之情。

谨以此书献给我们的妻子和孩子们。

Dan Stefanica

Radoš Radoičić

Tai-Ho Wang

二〇一七年，写于纽约

Acknowledgments

As professors in the Financial Engineering Masters Program at Baruch College, it has been a privilege to have the opportunity to contribute to the early career success of so many talented students and alumni. The strong community that developed around the Baruch MFE program is the reward educators truly dream of, and, for us, a reality that is an inspiration. This book would have not been possible without our Baruch MFE community – we are grateful to everyone who is part of it.

A special debt of gratitude is owed to our alumni Jun Hua (Baruch MFE'14) and Yu Gan (Baruch MFE'14) whose tireless work and dedication made the bilingual version of the book a reality. They both brought to the translation work a remarkable attention to details and the ambition to deliver the best possible translation - a reflection of the skills that are helping them forge very successful careers in industry!

The art for the book cover is, as always, due to the professional help of our alum and friend Max Rumyantsev.

Several students spearheaded the proofreading effort of the bilingual version, and their help was greatly appreciated: Huiyou (Francis) Chen, Zhihao (Alex) Chen, Mingyang (Corbin) Guan, Zijun (Alfred) Huang, Weihao (Vincent) Li, Sheng (Matthew) Lin, Yuchen (Derek) Qi,

Zilun Shen, Xinlu Xiao, Gongshun (Gordon) Yin, Mengqi (Kiki) Zhang, Mengyu (Kristin) Zhang, and ShengQuan Zhou.

We are also thankful to our alumni who helped proofreading the original edition of the book: Alejandro Cañete first and foremost, Yu Gan and Jun Hua once more, and Alireza Kashef, Yi Bill Lu, Svetlana Rafailova, Fubo Shi, Yujia Helen Sun, Jun Charlotte Wang, Peng Wu, Yanzhu Wendy Wu, Yongyi Ivan Ye, He Hillary Zhao, and Wenyi Zhou.

It is the tremendous support and the understanding of our families that made this book possible, and we are forever in debt to them. Thank you, thank you, and thank you!

This book is dedicated to our wives and to our children, from all our hearts.

Dan Stefanica

Radoš Radoičić

Tai-Ho Wang

New York, 2013

Chapter 1

小试牛刀：精选十道题.

1. 两个认沽期权的标的物和到期日均相同，行权价格分别为$30和$20. 如果它们的售价为$6和$4，是否存在套利机会？

2. 2^{29}是一个9位数，9个数字各不相同. 请问不在其中的那个数字是几？(不必算出2^{29})

3. W_t是维纳过程，令

$$X_t = \int_0^t W_\tau d\tau.$$

X_t的分布是怎样的？X_t是鞅吗？

4. 爱丽丝和鲍勃站在一条线段的两端. 鲍勃朝爱丽丝的方向放出50只蚂蚁，爱丽丝朝鲍勃的方向放出20只蚂蚁，蚂蚁一只接着一只地朝对方爬去. 两只蚂蚁相遇时，它们都会掉头继续爬行. 请问最终有多少只蚂蚁爬到鲍勃处，多少只爬到爱丽丝处？一共发生了多少次相遇？

5. 找到ρ的取值范围使得

$$\begin{pmatrix} 1 & 0.6 & -0.3 \\ 0.6 & 1 & \rho \\ -0.3 & \rho & 1 \end{pmatrix}$$

是一个相关系数矩阵.

6. 需要生成多少个相互独立的$[0,1]$上的均匀分布随机变量, 才能保证其中至少有一个在0.70和0.72之间的概率为95%?

7. 证明标准正态分布的概率密度函数的积分为1.

8. 假设地球是一个完美的球形, 你站在地球的表面上. 你往南走了1英里, 往东走了1英里, 又往北走了1英里. 你惊喜地发现自己回到了初始位置. 除了北极以外, 你的初始位置还可能是哪里?

9. 求解Ornstein-Uhlenbeck SDE

$$dr_t \ = \ \lambda(\theta - r_t)dt \ + \ \sigma dW_t,$$

(这个随机微分方程被用在Vasicek模型中.)

10. 编写一个C++函数, 计算第n个斐波那契数.

参考答案

题目 1. 两个认沽期权的标的物和到期日均相同, 行权价格分别为\$30和\$20. 如果它们的售价为\$6和\$4, 是否存在套利机会?

参考答案: 因为行权价格为\$0的认沽期权价值为\$0，我们实际上知道三个不同行权价的期权售价，即

$$P(30) = 6;\ P(20) = 4;\ P(0) = 0,$$

其中$P(K)$为行权价为K的认沽期权的价格.

在平面$(K, P(K))$上，这3个认沽期权分别对应 $(30, 6)$，$(20, 4)$，$(0, 0)$这三个点.我们注意到这三点共线，对应的直线方程为 $P(K) = \frac{2}{3}K$.

我们知道认沽期权的价格应当是关于行权价的严格凹函数，当这一性质被破坏时，就出现了套利机会.

这里的套利机会来源于行权价为20的认沽期权定价过高.根据“低买高卖”的策略，我们应当买入$\frac{2}{3}$份行权价为30的认沽期权，同时卖出1份行权价为20的认沽期权. 为了避免处理小数，我们建立以下投资组合:

- 买入2份行权价格为30的认沽期权;
- 卖出3份行权价格为20的认沽期权.

建立这份投资组合的成本为0，因为由卖出3份行权价格为20的认沽期权和买入2份行权价格为30的认沽期权产生的现金流总和为0.

$$3 \cdot \$4 - 2 \cdot \$6 \ = \ \$0.$$

在期权的到期日T，这份投资组合的价值为

$$V(T) \ = \ 2\max(30 - S(T), 0) - 3\max(20 - S(T), 0).$$

注意到对于任意的标的物价格$S(T)$:

如果$S(T) \geq 30$，则两种期权都没有价值，因此$V(T) = 0$.

如果$20 \le S(T) < 30$，则

$$V(T) = 2(30 - S(T)) > 0.$$

如果$0 < S(T) < 20$，则

$$\begin{aligned} V(T) &= 2(30 - S(T)) - 3(20 - S(T)) \\ &= S(T) \\ &> 0. \end{aligned}$$

换言之，我们利用这个套利机会无成本地建立了一份投资组合. 无论在到期时标的物的价格$S(T)$是多少，这份投资组合的回报都非负，并且当$0 < S(T) < 30$ 时, 回报严格为正. □

题目 2. 2^{29}是一个9位数, 9个数字各不相同. 请问不在其中的那个数字是几?(不必算出2^{29})

参考答案: 对于任意正整数n，用$D(n)$表示n各位数的和. 我们知道一个数和它各位数之和的差能被9整除，即

$$9 \mid n - D(n);$$

关于此式的证明，请参考脚注[1].

因此，对于$n = 2^{29}$，我们得到

$$9 \mid 2^{29} - D\left(2^{29}\right). \tag{1.1}$$

[1] 如果n的各位数为a_k, a_{k-1}, $\ldots a_1$, a_0 (从左到右)，则

$$\begin{aligned} n &= a_k \cdot 10^k + a_{k-1} \cdot 10^{k-1} + \ldots a_1 \cdot 10 + a_0; \\ D(n) &= a_k + a_{k-1} + \ldots + a_1 + a_0. \end{aligned}$$

从而，

$$n - D(n) = \sum_{i=0}^{k} a_i \cdot (10^i - 1).$$

由于$10^i - 1$是一个各位数都为9的i位数，所以对于任意$i = 1 : k$，$9 \mid 10^i - 1$，因此$9 \mid n - D(n)$.

我们已知2^{29}是九位数，而且九个数字都不同.用x表示缺少的数字，则

$$D\left(2^{29}\right)=\left(\sum_{j=0}^{9} j\right)-x=45-x. \tag{1.2}$$

由(1.1)和(1.2)，可以得到

$$9 \mid 2^{29}-(45-x). \tag{1.3}$$

注意到

$$\begin{aligned}
2^{29} &= 2^5\cdot(2^6)^4 = 2^5\cdot 64^4 \\
&= 2^5\cdot(63+1)^4 \\
&= 2^5\cdot(63\cdot k+1) \\
&= 2^5\cdot 63\cdot k+2^5,
\end{aligned} \tag{1.4}$$

其中k是正整数. [2]

根据(1.4)，我们发现

$$2^{29}-2^5 = 63\cdot 2^5\cdot k,$$

从而

$$9 \mid 2^{29}-2^5. \tag{1.5}$$

由(1.3)和(1.5)，可以得到

$$\begin{aligned}
9 \quad \mid \quad & (2^{29}-2^5) - (2^{29}-(45-x)) \\
&=(45-x)-2^5 \\
&=13-x.
\end{aligned}$$

---

[2] 容易得到

$$\begin{aligned}
(63+1)^4 &= 63^4+4\cdot 63^3+6\cdot 63^2+4\cdot 63+1 \\
&= 63\cdot(63^3+4\cdot 63^2+6\cdot 63+4)+1 \\
&= 63\cdot k+1,
\end{aligned}$$

其中 $k=63^3+4\cdot 63^2+6\cdot 63+4$.

由于$9 \mid 13-x$，且x是一位数，我们得到结论$x=4$. 换言之，2^{29}中缺少的数字就是4.

事实上，$2^{29}=536\,870\,912$，正好是九位数，每个数字都不同，且4没有在其中出现. □

题目 3. W_t是维纳过程, 令

$$X_t = \int_0^t W_\tau d\tau. \tag{1.6}$$

X_t的分布是怎样的?X_t是鞅吗?

参考答案: 注意到我们可以把 (1.6) 写作微分形式

$$dX_t = W_t dt = W_t dt + 0\,dW_t.$$

所以， X_t是只有漂移项W_t的扩散过程，因此不是鞅.

我们采用分部积分来计算X_t的分布；在5.6节中我们会展示另一种解法.

利用分部积分，我们得到

$$\begin{aligned} X_t &= \int_0^t W_\tau d\tau \\ &= tW_t - \int_0^t \tau dW_\tau \\ &= t\int_0^t dW_\tau - \int_0^t \tau dW_\tau \\ &= \int_0^t (t-\tau) dW_\tau. \end{aligned}$$

我们知道，如果$f(t)$是平方可积的确定性函数，则随机积分$\int_0^t f(\tau)dW_\tau$服从均值为0，方差为$\int_0^t |f(\tau)|^2 d\tau$的正态分布，即，

$$\int_0^t f(\tau)dW_\tau \sim N\left(0, \int_0^t |f(\tau)|^2 d\tau\right).$$

因此

$$\begin{aligned} X_t &= \int_0^t (t-\tau) dW_\tau \\ &\sim N\left(0, \int_0^t (t-\tau)^2 \, d\tau\right) \\ &= N\left(0, \frac{t^3}{3}\right). \end{aligned}$$

我们得到结论，X_t是均值为0，方差为$\frac{t^3}{3}$的正态随机变量. □

题目 4. 爱丽丝和鲍勃站在一条线段的两端. 鲍勃朝爱丽丝的方向放出50只蚂蚁, 爱丽丝朝鲍勃的方向放出20只蚂蚁, 蚂蚁一只接着一只地朝对方爬去. 两只蚂蚁相遇时, 它们都会掉头继续爬行. 请问最终有多少只蚂蚁爬到鲍勃处, 多少只爬到爱丽丝处? 一共发生了多少次相遇?

参考答案: 想象两只蚂蚁相遇时，它们交换身份.这样，即使两只蚂蚁相遇，它们仍然向相反方向爬去.这样，一定有20只蚂蚁爬到鲍勃处， 50只蚂蚁爬到爱丽丝处.

为了计算蚂蚁相遇的次数，我们假定每只蚂蚁都带着一条讯息.换言之，鲍勃向爱丽丝发出50条信息，由每只蚂蚁携带一条.同样的，爱丽丝向鲍勃发出20条信息，由每只蚂蚁携带一条.进一步假设两只蚂蚁相遇时交换信息.这样，每条信息都总是向前传递.爱丽丝发出的每条消息都要经过50次相遇，鲍勃的每条信息则要经过20次相遇.总共的相遇次数是50乘以20，即1000次. □

题目 5. 找到ρ的取值范围使得

$$\begin{pmatrix} 1 & 0.6 & -0.3 \\ 0.6 & 1 & \rho \\ -0.3 & \rho & 1 \end{pmatrix}$$

是一个相关系数矩阵.

参考答案: 对角线为1的对称矩阵是相关系数矩阵, 当且仅当其为半正定矩阵. 因此, 我们要找到ρ的取值范围使以下矩阵半正定.

$$\Omega = \begin{pmatrix} 1 & 0.6 & -0.3 \\ 0.6 & 1 & \rho \\ -0.3 & \rho & 1 \end{pmatrix} \tag{1.7}$$

这里我们给出一种用到Sylvester条件的解法. 另外两种解法(一种用到Cholesky分解, 一种基于半正定矩阵的定义)将会在章节5.2中给出.

Sylvester条件指一个矩阵半正定当且仅当其所有主子式大于等于0. 主子式指的行号与列号相同的子方阵的行列式.

式(1.7)中的矩阵Ω有如下主子式:

$$\det(1) = 1; \quad \det(1) = 1; \quad \det(1) = 1;$$

$$\begin{aligned} \det\begin{pmatrix} 1 & 0.6 \\ 0.6 & 1 \end{pmatrix} &= 0.64; \\ \det\begin{pmatrix} 1 & -0.3 \\ -0.3 & 1 \end{pmatrix} &= 0.91; \\ \det\begin{pmatrix} 1 & \rho \\ \rho & 1 \end{pmatrix} &= 1-\rho^2; \end{aligned}$$

$$\begin{aligned} \det(\Omega) &= 1 - 0.36\rho - 0.09 - 0.36 - \rho^2 \\ &= 0.55 - 0.36\rho - \rho^2. \end{aligned}$$

因此, 根据Sylvester条件, Ω半正定当且仅当

$$\begin{aligned} 1-\rho^2 &\geq 0; \\ 0.55 - 0.36\rho - \rho^2 &\geq 0, \end{aligned}$$

其等价于 $-1 \leq \rho \leq 1$ 以及

$$\rho^2 + 0.36\rho - 0.55 \leq 0. \tag{1.8}$$

因为式(1.8)的根为-0.9432和0.5832, 我们得到 Ω是相关系数矩阵当且仅当

$$-0.9432 \leq \rho \leq 0.5832. \quad \square \tag{1.9}$$

题目 6. 需要生成多少个相互独立的$[0,1]$上的均匀分布随机变量, 才能保证其中至少有一个在0.70和0.72之间的概率为95%?

参考答案: 记N为使下式成立的最小的随机变量个数

$$P(\text{至少有一个随机变量在}[0.70, 0.72]\text{里}) \geq 0.95. \tag{1.10}$$

一个在$[0,1]$间均匀分布的随机变量不在区间$[0.70, 0.72]$里的概率为0.98. 因此, N个独立的随机变量均不在$[0.70, 0.72]$里的概率是0.98^N, 即,

$$P(\text{没有随机变量在}[0.70, 0.72]\text{里}) = 0.98^N.$$

注意到

$$\begin{aligned} & P(\text{至少有一个随机变量在}[0.70, 0.72]\text{里}) \\ = \; & 1 - P(\text{没有随机变量在}[0.70, 0.72]\text{里}) \\ = \; & 1 - (0.98)^N. \end{aligned} \tag{1.11}$$

由式(1.10)和式(1.11), 可得N是使下式成立的最小整数

$$1 - (0.98)^N \geq 0.95,$$

其等价于

$$\begin{aligned} & (0.98)^N \leq 0.05 \\ \iff \; & N \ln(0.98) \leq \ln(0.05) \\ \iff \; & N \geq \frac{\ln(0.05)}{\ln(0.98)} \approx 148.28 \\ \iff \; & N = 149. \end{aligned}$$

综上, 为了保证至少有一个变量在0.70和0.72之间的概率为95%, 至少要生成149个[0, 1]间均匀分布的随机变量.
□

题目 7. 证明标准正态分布的概率密度函数的积分为1.

参考答案: 标准正态分布的概率密度函数为 $\frac{1}{\sqrt{2\pi}}\,e^{-\frac{t^2}{2}}$. 我们想要证明

$$\frac{1}{\sqrt{2\pi}}\int_{-\infty}^{\infty} e^{-\frac{t^2}{2}}\,dt \;=\; 1,$$

做变量替换$t = \sqrt{2}x$, 上式可以写作

$$I \;=\; \int_{-\infty}^{\infty} e^{-x^2}\,dx \;=\; \sqrt{\pi}. \tag{1.12}$$

接下来我们用极坐标系来证明式(1.12). 因为x是积分变量, 我们可以把I写成关于另一个积分变量y的积分 $I = \int_{-\infty}^{\infty} e^{-y^2}\,dy$.

所以,[3]

$$\begin{aligned} I^2 &= \left(\int_{-\infty}^{\infty} e^{-x^2}\,dx\right)\cdot\left(\int_{-\infty}^{\infty} e^{-y^2}\,dy\right) \\ &= \int_{-\infty}^{\infty}\int_{-\infty}^{\infty} e^{-x^2}e^{-y^2}\,dxdy \qquad (1.13) \\ &= \int_{-\infty}^{\infty}\int_{-\infty}^{\infty} e^{-(x^2+y^2)}\,dxdy. \end{aligned}$$

对最后一个积分使用极坐标系变换 $x = r\cos\theta$和$y = r\sin\theta$,

---

[3]注意到等式(1.13)的严格推导要用到Fubini定理; 这个技术步骤几乎不会被面试官问到.

其中$r \in [0, \infty)$, $\theta \in [0, 2\pi)$. 因为$dxdy = rd\theta dr$, 我们得到

$$
\begin{aligned}
I^2 &= \int_{-\infty}^{\infty}\int_{-\infty}^{\infty} e^{-(x^2+y^2)}\, dxdy \\
&= \int_0^{\infty}\int_0^{2\pi} r\, e^{-\left(r^2\cos^2\theta + r^2\sin^2\theta\right)}\, d\theta dr \\
&= \int_0^{\infty}\int_0^{2\pi} r\, e^{-r^2}\, d\theta dr \qquad (1.14) \\
&= \int_0^{\infty} 2\pi\, r\, e^{-r^2}\, dr \\
&= 2\pi \lim_{t\to\infty} \int_0^{t} r\, e^{-r^2}\, dr \\
&= 2\pi \lim_{t\to\infty} \left(-\frac{1}{2}e^{-r^2}\right)\Big|_0^t \\
&= \pi;
\end{aligned}
$$

注意到式(1.14)用到了等式 $\cos^2\theta + \sin^2\theta = 1$对任意实数$\theta$成立.

因为$I > 0$, 所以$I = \sqrt{\pi}$, 我们证明了式(1.12). □

题目 8. 假设地球是一个完美的球形, 你站在地球的表面上. 你往南走了1英里, 往东走了1英里, 又往北走了1英里. 你惊喜地发现自己回到了初始位置. 除了北极以外, 你的初始位置还可能是哪里?

参考答案: 除北极之外, 满足这个条件的位置还有无数个.

在地球表面上, 南极附近有一圈周长为一英里的纬线. 从这条纬线上的任意一点P出发, 往东(或西)走一英里, 你将回到点P. 如果你从点P以北一英里处出发, 往南一英里走到点P. 接着往东走一英里回到点P(绕纬线走一圈), 最后往北一英里回到初始位置.

考虑到点P为这条特殊纬线上的任意一点, 初始位置为点P以北一英里, 我们有无数个满足条件的初始位置.

事实上, 这样的特殊纬线也有无数条. 南极附近任意周长为$1/k$英里(k为正整数)的纬线都可以. 从这样的纬线以北一英里出发, 往南一英里走到特殊纬线上的一个点, 往东一

英里回到这个点(绕纬线走k圈), 最终往北一英里回到初始位置. □

题目 9. 求解Ornstein-Uhlenbeck SDE

$$dr_t \;=\; \lambda(\theta - r_t)dt \;+\; \sigma dW_t, \tag{1.15}$$

(这个随机微分方程被用在Vasicek模型中.)

参考答案: 我们把式(1.15)写作

$$dr_t + \lambda r_t dt \;=\; \lambda\theta dt + \sigma dW_t. \tag{1.16}$$

在式(1.16)的等号两边同时乘以$e^{\lambda t}$, 我们得到

$$e^{\lambda t}dr_t + \lambda e^{\lambda t}r_t dt \;=\; \lambda\theta e^{\lambda t}dt + \sigma e^{\lambda t}dW_t,$$

其等价于

$$d\left(e^{\lambda t}r_t\right) \;=\; \lambda\theta e^{\lambda t}dt + \sigma e^{\lambda t}dW_t. \tag{1.17}$$

对式(1.17)从0到t积分,有

$$\begin{aligned} e^{\lambda t}r_t - r_0 &= \lambda\theta\int_0^t e^{\lambda s}ds + \sigma\int_0^t e^{\lambda s}dW_s \\ &= \theta\left(e^{\lambda t} - 1\right) + \sigma\int_0^t e^{\lambda s}dW_s\,. \end{aligned}$$

所以Ornstein-Uhlenbeck SDE的解为

$$\begin{aligned} r_t &= e^{-\lambda t}r_0 + e^{-\lambda t}\theta\left(e^{\lambda t} - 1\right) + \sigma e^{-\lambda t}\int_0^t e^{\lambda s}dW_s \\ &= e^{-\lambda t}r_0 + \theta\left(1 - e^{-\lambda t}\right) + \sigma\int_0^t e^{-\lambda(t-s)}dW_s. \end{aligned}$$

注意到无论起始位置r_0在哪, 随机过程r_t都会朝θ均值回归. 这是因为非随机函数$f(s)$的随机积分$\int_0^t f(s)dW_s$的期望值为0, 故而

$$E\left[\int_0^t e^{-\lambda(t-s)}dW_s\right] = 0,$$

所以

$$E[r_t] = e^{-\lambda t} r_0 + \theta \left(1 - e^{-\lambda t}\right).$$

所以,

$$\lim_{t \to \infty} E[r_t] = \theta. \quad \square$$

题目 10. 编写一个C++函数, 计算第n个斐波那契数.

参考答案: 斐波那契数列$(F_n)_{n \geq 0}$由以下递归式定义:

$$F_{n+2} = F_{n+1} + F_n, \ \forall n \geq 0,$$

同时$F_0 = 0$, $F_1 = 1$.

```
//recursive implementation
int fib(int n) {
    if (n == 0 || n == 1) return n;
    else {
        return fib(n-1) + fib(n-2);
    }
}

//iterative implementation
int fib(int n ){
    if (n == 0 || n == 1) return n;
    int prev = 0, last = 1, temp;
    for (int i = 2; i <= n; ++i) {
        temp = last;
        last = prev + last;
        prev = temp;
    }
    return last;
}

//tail recursive implementation
int fib(int n, int last = 1, int prev = 0)
{
```

```
    if (n == 0) return prev;
    if (n == 1) return last;
        return fib(n-1, last+prev, last);
}
```

Chapter 2

First Look: Ten Questions.

1. Put options with strikes 30 and 20 on the same underlying asset and with the same maturity are trading for \$6 and \$4, respectively. Can you find an arbitrage?

2. The number 2^{29} has 9 digits, all different. Without computing 2^{29}, find the missing digit.

3. Let W_t be a Wiener process, and let

$$X_t = \int_0^t W_\tau d\tau.$$

What is the distribution of X_t? Is X_t a martingale?

4. Alice and Bob stand at opposite ends of a straight line segment. Bob sends 50 ants towards Alice, one after another. Alice sends 20 ants towards Bob. All ants travel along the straight line segment. Whenever two ants collide, they simply bounce back and start traveling in the opposite direction. How many

ants reach Bob and how many ants reach Alice? How many ant collisions take place?

5. Find all the values of ρ such that

$$\begin{pmatrix} 1 & 0.6 & -0.3 \\ 0.6 & 1 & \rho \\ -0.3 & \rho & 1 \end{pmatrix}$$

is a correlation matrix.

6. How many independent random variables uniformly distributed on $[0, 1]$ should you generate to ensure that there is at least one between 0.70 and 0.72 with probability 95%?

7. Show that the probability density function of the standard normal integrates to 1.

8. Assume the Earth is perfectly spherical and you are standing somewhere on its surface. You travel exactly 1 mile south, then 1 mile east, then 1 mile north. Surprisingly, you find yourself back at the starting point. If you are not at the North Pole, where can you possibly be?!

9. Solve the Ornstein-Uhlenbeck SDE

$$dr_t = \lambda(\theta - r_t)dt + \sigma dW_t,$$

which is used, e.g., in the Vasicek model for interest rates.

10. Write a C++ function that computes the n-th Fibonacci number.

Solutions

Question 1. Put options with strikes 30 and 20 on the same underlying asset and with the same maturity are trading for \$6 and \$4, respectively. Can you find an arbitrage?

Answer: Since the value of a put option with strike 0 is \$0, we in fact know the prices of put options with three different strikes, i.e.,

$$P(30) = 6; \; P(20) = 4; \; P(0) = 0,$$

where $P(K)$ denotes the value of a put option with strike K.

In the plane $(K, P(K))$, these option values correspond to the points $(30, 6)$, $(20, 4)$, and $(0, 0)$, which are on the line $P(K) = \frac{2}{3}K$.

This contradicts the fact that put options are strictly convex functions of strike price, and creates an arbitrage opportunity.

The arbitrage comes from the fact that the put with strike 20 is overpriced. Using a "buy low, sell high" strategy, we could buy (i.e., go long) $\frac{2}{3}$ put options with strike 30, and sell (i.e., go short) 1 put option with strike 20. To avoid fractions, we set up the following portfolio:

- long 2 puts with strike 30;
- short 3 puts with strike 20.

This portfolio is set up at no initial cost, since the cash flow generated by selling 3 puts with strike 20 and buying 2 puts with strike 30 is \$0:

$$3 \cdot \$4 - 2 \cdot \$6 \;=\; \$0.$$

At the maturity T of the options, the value of the portfolio is

$$V(T) \;=\; 2\max(30 - S(T), 0) - 3\max(20 - S(T), 0).$$

Note that $V(T)$ is nonnegative for any value $S(T)$ of the underlying asset:

If $S(T) \geq 30$, then both put options expire worthless, and $V(T) = 0$.

If $20 \leq S(T) < 30$, then

$$V(T) \;=\; 2(30 - S(T)) \;>\; 0.$$

If $0 < S(T) < 20$, then

$$\begin{aligned} V(T) &= 2(30 - S(T)) - 3(20 - S(T)) \\ &= S(T) \\ &> 0. \end{aligned}$$

In other words, we took advantage of the existing arbitrage opportunity by setting up, at no initial cost, a portfolio with nonnegative payoff at T regardless of the price $S(T)$ of the underlying asset, and with a strictly positive payoff if $0 < S(T) < 30$. □

Question 2. The number 2^{29} has 9 digits, all different. Without computing 2^{29}, find the missing digit.

Answer: For any positive integer n, denote by $D(n)$ the sum of the digits of n. Recall that the difference between a number and the sum of its digits is divisible by 9, i.e.,

$$9 \mid n - D(n);$$

see the footnote below[1] for details.

---

[1]If the digits of n are $a_k, a_{k-1}, \ldots a_1, a_0$ (from left to right), then

$$\begin{aligned} n &= a_k \cdot 10^k + a_{k-1} \cdot 10^{k-1} + \ldots a_1 \cdot 10 + a_0; \\ D(n) &= a_k + a_{k-1} + \ldots + a_1 + a_0. \end{aligned}$$

Hence,

$$n - D(n) = \sum_{i=0}^{k} a_i \cdot (10^i - 1).$$

Thus, for $n = 2^{29}$, it follows that

$$9 \mid 2^{29} - D\left(2^{29}\right). \tag{2.1}$$

We are given that 2^{29} has 9 digits, and that all 9 digits are different. Denote by x the missing digit. Then,

$$D\left(2^{29}\right) = \left(\sum_{j=0}^{9} j\right) - x = 45 - x. \tag{2.2}$$

From (2.1) and (2.2), it follows that

$$9 \mid 2^{29} - (45 - x). \tag{2.3}$$

Note that

$$\begin{aligned} 2^{29} &= 2^5 \cdot (2^6)^4 = 2^5 \cdot 64^4 \\ &= 2^5 \cdot (63+1)^4 \\ &= 2^5 \cdot (63 \cdot k + 1) \\ &= 2^5 \cdot 63 \cdot k + 2^5, \end{aligned} \tag{2.4}$$

where k is a positive integer.[2]

From (2.4), we find that

$$2^{29} - 2^5 = 63 \cdot 2^5 \cdot k,$$

and therefore

$$9 \mid 2^{29} - 2^5. \tag{2.5}$$

Since $10^i - 1$ is an i-digit number with all digits equal to 9, it follows that $9 \mid 10^i - 1$, for all $i = 1 : k$, and therefore $9 \mid n - D(n)$.

[2] It is easy to see that

$$\begin{aligned} (63+1)^4 &= 63^4 + 4 \cdot 63^3 + 6 \cdot 63^2 + 4 \cdot 63 + 1 \\ &= 63 \cdot (63^3 + 4 \cdot 63^2 + 6 \cdot 63 + 4) + 1 \\ &= 63 \cdot k + 1, \end{aligned}$$

where $k = 63^3 + 4 \cdot 63^2 + 6 \cdot 63 + 4$.

The probability that a random variable uniformly distributed on $[0,1]$ is not in the interval $[0.70, 0.72]$ is 0.98. Thus, the probability that none of the N independent variables are in $[0.70, 0.72]$ is 0.98^N, i.e.,

$$P(\text{no r.v. in } [0.70, 0.72]) \;=\; 0.98^N.$$

Note that

$$\begin{aligned} & P(\text{at least one r.v. in } [0.70, 0.72]) \\ = \;& 1 - P(\text{no r.v. in } [0.70, 0.72]) \\ = \;& 1 - (0.98)^N. \end{aligned} \qquad (2.11)$$

From (2.10) and (2.11), we find that N is the smallest integer such that

$$1 - (0.98)^N \;\geq\; 0.95,$$

which is equivalent to

$$\begin{aligned} & (0.98)^N \leq 0.05 \\ \Longleftrightarrow \quad & N \ln(0.98) \leq \ln(0.05) \\ \Longleftrightarrow \quad & N \geq \frac{\ln(0.05)}{\ln(0.98)} \approx 148.28 \\ \Longleftrightarrow \quad & N = 149. \end{aligned}$$

We conclude that at least 149 uniform random variables on $[0,1]$ must be generated in order to have 95% confidence that at least one of the random variables is between 0.70 and 0.72. □

Question 7. Show that the probability density function of the standard normal integrates to 1.

Answer: The probability density function of the standard normal variable is $\frac{1}{\sqrt{2\pi}}\, e^{-\frac{t^2}{2}}$. We want to show that

$$\frac{1}{\sqrt{2\pi}} \int_{-\infty}^{\infty} e^{-\frac{t^2}{2}}\, dt \;=\; 1,$$

which, using the substitution $t = \sqrt{2}x$, can be written as

$$I = \int_{-\infty}^{\infty} e^{-x^2}\, dx = \sqrt{\pi}. \qquad (2.12)$$

We prove (2.12) by using polar coordinates. Since x is just an integrating variable, we can also write the integral I in terms of another integrating variable, denoted by y, as $I = \int_{-\infty}^{\infty} e^{-y^2}\, dy$.

Then,[3]

$$\begin{aligned} I^2 &= \left(\int_{-\infty}^{\infty} e^{-x^2}\, dx\right) \cdot \left(\int_{-\infty}^{\infty} e^{-y^2}\, dy\right) \\ &= \int_{-\infty}^{\infty}\int_{-\infty}^{\infty} e^{-x^2} e^{-y^2}\, dxdy \qquad (2.13) \\ &= \int_{-\infty}^{\infty}\int_{-\infty}^{\infty} e^{-(x^2+y^2)}\, dxdy. \end{aligned}$$

We use the polar coordinates transformation $x = r\cos\theta$ and $y = r\sin\theta$, with $r \in [0, \infty)$ and $\theta \in [0, 2\pi)$, to evaluate the last integral. Since $dxdy = rd\theta dr$, we obtain

[3]Note that Fubini's theorem is needed for a rigorous derivation of the equality (2.13); this technical step is rarely required by the interviewer.

that

$$
\begin{aligned}
I^2 &= \int_{-\infty}^{\infty}\int_{-\infty}^{\infty} e^{-(x^2+y^2)}\,dxdy \\
&= \int_0^{\infty}\int_0^{2\pi} r\,e^{-\left(r^2\cos^2\theta + r^2\sin^2\theta\right)}\,d\theta dr \\
&= \int_0^{\infty}\int_0^{2\pi} r\,e^{-r^2}\,d\theta dr \qquad (2.14) \\
&= \int_0^{\infty} 2\pi\,r\,e^{-r^2}\,dr \\
&= 2\pi\,\lim_{t\to\infty}\int_0^t r\,e^{-r^2}\,dr \\
&= 2\pi\,\lim_{t\to\infty}\left(-\frac{1}{2}e^{-r^2}\right)\Big|_0^t \\
&= \pi;
\end{aligned}
$$

note that (2.14) follows from the equality $\cos^2\theta + \sin^2\theta = 1$ for any real number θ.

Since $I > 0$, $I = \sqrt{\pi}$, which is what we wanted to prove; see (2.12). □

Question 8. Assume the Earth is perfectly spherical and you are standing somewhere on its surface. You travel exactly 1 mile south, then 1 mile east, then 1 mile north. Surprisingly, you find yourself back at the starting point. If you are not at the North Pole, where can you possibly be?!

Answer: There are infinitely many locations, aside from the North Pole, that have this property.

Somewhere near the South Pole, there is a latitude that has a circumference of one mile. In other words, if you are at this latitude and start walking east (or west), in one mile you will be back exactly where you started from. If you instead start at some point one mile north of this latitude, your journey will take you one mile south to this special latitude, then one mile east "around the globe" and finally one mile north right back to wherever you started from. Moreover, there are infinitely many points on the Earth that are one mile north of this special latitude, where you could start your journey and eventually end up exactly where you started.

We are still not finished! There are infinitely many special latitudes as well; namely, you could start at any point one mile north of the latitude that has a circumference of $1/k$ miles, where k is a positive integer. Your journey will take you one mile south to this special latitude, then one mile east looping "around the globe" k times, and finally one mile north right back to where you started from. □

Question 9. Solve the Ornstein-Uhlenbeck SDE

$$dr_t = \lambda(\theta - r_t)dt + \sigma dW_t, \tag{2.15}$$

which is used, e.g., in the Vasicek model for interest rates.

Answer: We can rewrite (2.15) as

$$dr_t + \lambda r_t dt = \lambda\theta dt + \sigma dW_t. \tag{2.16}$$

By multiplying (2.16) on both sides by the integrating factor $e^{\lambda t}$, we obtain that

$$e^{\lambda t} dr_t + \lambda e^{\lambda t} r_t dt = \lambda\theta e^{\lambda t} dt + \sigma e^{\lambda t} dW_t,$$

which is equivalent to

$$d\left(e^{\lambda t} r_t\right) = \lambda\theta e^{\lambda t} dt + \sigma e^{\lambda t} dW_t. \tag{2.17}$$

By integrating (2.17) from 0 to t, it follows that

$$\begin{aligned} e^{\lambda t} r_t - r_0 &= \lambda\theta \int_0^t e^{\lambda s} ds + \sigma \int_0^t e^{\lambda s} dW_s \\ &= \theta\left(e^{\lambda t} - 1\right) + \sigma \int_0^t e^{\lambda s} dW_s. \end{aligned}$$

8. Delta的取值范围是什么?

9. 平值认购期权和平值认沽期权的Delta分别是多少?

10. 什么是买卖权平价关系(Put-Call Parity)? 如何证明其成立?

11. 证明欧式认购期权在平值时时间价值最大.

12. 什么是隐含波动率(implied volatility)?
    波动率微笑(volatility smile)?
    波动率偏离(volatility skew)?

13. 什么是期权的$Gamma$?为什么$Gamma$越小越容易控制风险?为什么欧式期权的$Gamma$大于0?

14. 欧式认购期权的价格与欧式认沽期权的价格在什么情况下相同? (上述认购与认沽期权的行权价格,标的物和到期日均相同) 如何直观地理解本题的结论?

15. 如果一个标的物的六个月波动率为30%,那么它两年的波动率为多少?

16. 利率掉期(interest rate swap)是如何定价的?

17. 如果债券收益率上涨10个基准点,10年无息债券(zero coupon bond)的价格变化多少?

18. 一个久期(duration)为3.5年的5年期债券的价格为102.如果收益率(yield) 下降50个基点,这个债券的价格变化多少?

19. 什么是远期合约(forward contract)? 什么是远期价格(forward price)?

20. 国债期货合约(treasury futures contract)的远期价格应当如何计算? 大宗商品期货合约(commodities futures contract)的远期价格又该怎样计算呢?

21. 什么是欧洲美元期货(Eurodollar futures)?

22. 远期合约(forward contract)和期货合约(futures contract)的区别有哪些?

23. 如果一个投资组合5天的98%VAR为1千万,那么10天的99%VAR为多少?

3.5 蒙特卡洛模拟.数理方法.

1. 如何用蒙特卡洛计算π?计算结果的标准差为多少?

2. 产生独立标准正态分布样本的方法有哪些?

3. 如何用正态分布随机数生成器来生成符合几何布朗运动的股票价格走势?

4. 如何用12个在$[0, 1]$上的均匀分布的样本生成一个近似服从标准正态分布的样本?

5. 蒙特卡洛的收敛速度是多少?

6. 方差缩减(variance reduction)有哪些"奇技淫巧"?

7. 如何生成两个相关系数为ρ的正态分布样本?

8. 牛顿法的收敛阶数是多少?

9. 哪一种有限差分法(finite difference method)与三叉树模型(trinomial trees)相对应?

3.6 概率论. 随机微分.

1. 什么是指数分布? 指数分布的均值和方差是什么?

2. X和Y是独立的指数分布随机变量, 期望分别为6和8, 求Y大于X的概率.

3. 泊松分布的期望值和方差是什么?

4. 从单位圆内随机(等概率)选取一个点, 它到圆心的距离的期望是多少?

5. 考虑两个联合正态分布的随机变量X和Y, 期望均为0, 方差均为1. 如果$\mathrm{cov}(X,Y) = \frac{1}{\sqrt{2}}$, 那么条件概率$P(X > 0|Y < 0)$是多少?

6. 如果X和Y均为对数正态随机变量, 它们的乘积也是对数正态分布吗?

7. X是期望为μ, 方差为σ^2的正态随机变量, Φ是标准正态分布的累积分布函数. 求$Y = \Phi(X)$的期望值.

8. 什么是大数定律?

9. 什么是中心极限定理?

10. 什么是鞅(Martingale)?鞅和期权定价有什么联系?

11. 请解释用在伊藤引理推导过程中的假设$(dW_t)^2 = dt$.

12. W_t是维纳过程, 求$E[W_t W_s]$.

13. W_t是维纳过程, 求$\mathrm{var}(W_t + W_s)$.

14. W_t是一个维纳过程, 求

$$\int_0^t W_s \, dW_s \text{ 和 } E\left[\int_0^t W_s \, dW_s\right].$$

15. 计算随机变量的概率密度分布

$$X = \int_0^1 W_t dW_t.$$

16. W_t是维纳过程, 求如下随机积分的均值和方差.

$$\int_0^t W_s^2 dW_s.$$

17. W_t是维纳过程, 求如下随机积分的方差.

$$X = \int_0^1 \sqrt{t} e^{\frac{W_t^2}{8}} dW_t.$$

18. W_t是维纳过程, 求$E\left[e^{W_t}\right]$.

19. W_t是维纳过程, 求如下随机积分的方差.

$$\int_0^t s\,dW_s.$$

20. W_t是维纳过程, 令

$$X_t = \int_0^t W_\tau d\tau. \tag{3.1}$$

X_t的分布是怎样的?X_t是鞅吗?[2]

21. 什么是伊藤过程?

22. 什么是伊藤引理?

23. W_t是维纳过程, $X_t = W_t^2$是鞅吗?

24. W_t是维纳过程, $N_t = W_t^3 - 3tW_t$是鞅吗?

25. 什么是Girsanov定理?

26. 什么是鞅表示定理?其与期权定价和对冲有什么关系?

27. 求解

$$dY_t = Y_t dW_t, \tag{3.2}$$

其中W_t是个维纳过程.

---

[2] 本题的一种解法 (使用分部积分) 已经在第1章中给出.我们将在本章中讨论另外一种解法.

9. Compute
$$\int \frac{1}{1+x^2}\,dx.$$

10. Compute
$$\int x\ln(x)dx \quad \text{and} \quad \int xe^x\,dx.$$

11. Compute
$$\int x^n \ln(x)\,dx.$$

12. Compute
$$\int (\ln(x))^n\,dx.$$

13. Solve the ODE
$$y'' - 4y' + 4y = 1.$$

14. Find $f(x)$ such that
$$f'(x) = f(x)(1 - f(x)).$$

15. Derive the Black-Scholes PDE.

4.2 Covariance and correlation matrices. Linear algebra.

1. Show that any covariance matrix is symmetric positive semidefinite. Show that the same is true for correlation matrices.

2. Find the correlation matrix of three random variables with covariance matrix

$$\Sigma_X = \begin{pmatrix} 1 & 0.36 & -1.44 \\ 0.36 & 4 & 0.80 \\ -1.44 & 0.80 & 9 \end{pmatrix}.$$

3. Assume that all the entries of an $n \times n$ correlation matrix which are not on the main diagonal are equal to ρ. Find upper and lower bounds on the possible values of ρ.

4. How many eigenvalues does an $n \times n$ matrix with real entries have? How many eigenvectors?

5. Let

$$A = \begin{pmatrix} 2 & -2 \\ -2 & 5 \end{pmatrix}.$$

(i) Find a 2×2 matrix M such that $M^2 = A$;

(ii) Find a 2×2 matrix M such that $A = MM^t$.

6. The 2×2 matrix A has eigenvalues 2 and -3 with corresponding eigenvectors $\begin{pmatrix} 1 \\ 2 \end{pmatrix}$ and $\begin{pmatrix} -1 \\ 3 \end{pmatrix}$. If $v = \begin{pmatrix} 3 \\ 1 \end{pmatrix}$, find Av.

7. Let A and B be square matrices of the same size. Show that the traces of the matrices AB and BA are equal.

8. Can you find $n \times n$ matrices A and B such that
$$AB - BA = I_n,$$
where I_n is the identity matrix of size n?

9. A probability matrix is a matrix with nonnegative entries such that the sum of the entries in each row of the matrix is equal to 1. Show that the product of two probability matrices is a probability matrix.

10. Find all the values of ρ such that
$$\begin{pmatrix} 1 & 0.6 & -0.3 \\ 0.6 & 1 & \rho \\ -0.3 & \rho & 1 \end{pmatrix}$$
is a correlation matrix.[1]

[1]A solution to this question was given in Chapter 2 using Sylvester's criterion; two different solutions will be given herein.

4.3 Financial instruments: options, bonds, swaps, forwards, futures.

1. The prices of three put options with strikes 40, 50, and 70, but otherwise identical, are \$10, \$20, and \$30, respectively. Is there an arbitrage opportunity present? If yes, how can you make a riskless profit?

2. The price of a stock is \$50. In three months, it will either be \$47 or \$52, with 50% probability. How much would you pay for an at–the–money put? Assume for simplicity that the stock pays no dividends and that interest rates are zero.

3. A stock worth \$50 today will be worth either \$60 or \$40 in three months, with equal probability. The value of a three months at–the–money put on this stock is \$4. Does the value of the three months ATM put increase or decrease, and by how much, if the probability of the stock going up to \$60 were 75% and the probability of the stock going down to \$40 were 25%?

4. What is risk–neutral pricing?

5. Describe briefly how you arrive at the Black–Scholes formula.

6. How much should a three months at–the–money put on an asset with spot price \$40 and volatility 30%

be worth? Assume, for simplicity, that interest rates are zero and that the asset does not pay dividends.

7. If the price of a stock doubles in one day, by how much will the value of a call option on this stock change?

8. What are the smallest and largest values that Delta can take?

9. What is the Delta of an at–the–money call? What is the Delta of an at–the–money put?

10. What is the Put–Call parity? How do you prove it?

11. Show that the time value of a European call option is highest at–the–money.

12. What is implied volatility? What is a volatility smile? How about a volatility skew?

13. What is the Gamma of an option? Why is it preferable to have small Gamma? Why is the Gamma of plain vanilla options positive?

14. When are a European call and a European put worth the same? (The options are written on the same asset and have the same strike and maturity.) What is the intuition behind this result?

15. What is the two year volatility of an asset with 30% six months volatility?

16. How do you value an interest rate swap?

17. By how much will the price of a ten year zero coupon bond change if the yield increases by ten basis points?

18. A five year bond with 3.5 years duration is worth 102. What is the value of the bond if the yield decreases by fifty basis points?

19. What is a forward contract? What is the forward price?

20. What is the forward price for treasury futures contracts? What is the forward price for commodities futures contracts?

21. What is a Eurodollar futures contract?

22. What are the most important differences between forward contracts and futures contracts?

23. What is the ten–day 99% VaR of a portfolio with a five–day 98% VaR of $10 million?

4.4 C++. Data structures.

1. How do you declare an array?

2. How do you get the address of a variable?

3. How do you declare an array of pointers?

4. How do you declare a const pointer, a pointer to a const and a const pointer to a const?

5. How do you declare a dynamic array?

6. What is the general form for a function signature?

7. How do you pass-by-reference?

8. How do you pass a read only argument by reference?

9. What are the important differences between using a pointer and a reference?

10. How do you set a default value for a parameter?

11. How do you create a template function?

12. How do you declare a pointer to a function?

13. How do you prevent the compiler from doing an implicit conversion with your class?

14. Describe all the uses of the keyword `static` in C++.

15. Can a static member function be const?

16. C++ constructors support the initialization of member data via an initializer list. When is this preferable to initialization inside the body of the constructor?

17. What is a copy constructor, and how can the default copy constructor cause problems when you have pointer data members?

18. What is the output of the following code:

```
#include <iostream>
using namespace std;

class A
{
public:
    int * ptr;
    ~A()
    {
        delete(ptr);
    }
};
```

```
void foo(A object_input)
{
    ;
}

int main()
{
    A aa;
    aa.ptr = new int(2);
    foo(aa);
    cout<<(*aa.ptr)<<endl;
    return 0;
}
```

19. How do you overload an operator?

20. What are smart pointers?

21. What is encapsulation?

22. What is a polymorphism?

23. What is inheritance?

24. What is a virtual function? What is a pure virtual function and when do you use it?

25. Why are virtual functions use for destructors? Can they be used for constructors?

26. Write a function that computes the factorial of a positive integer.

27. Write a function that takes an array and returns the subarray with the largest sum.

28. Write a function that returns the prime factors of a positive integer.

29. Write a function that takes a 64-bit integer and swaps the bits at indices `i` and `j`.

30. Write a function that reverses a single linked list.

31. Write a function that takes a string and returns true if its parenthesis are balanced.

32. Write a function that returns the height of an arbitrary binary tree.

4.6 Probability. Stochastic calculus.

1. What is the exponential distribution? What are the mean and the variance of the exponential distribution?

2. If X and Y are independent exponential random variables with mean 6 and 8, respectively, what is the probability that Y is greater than X?

3. What are the expected value and the variance of the Poisson distribution?

4. A point is chosen uniformly from the unit disk. What is the expected value of the distance between the point and the center of the disk?

5. Consider two random variables X and Y with mean 0 and variance 1, and with joint normal distribution. If $\mathrm{cov}(X, Y) = \frac{1}{\sqrt{2}}$, what is the conditional probability $P(X > 0|Y < 0)$?

6. If X and Y are lognormal random variables, is their product XY lognormally distributed?

7. Let X be a normal random variable with mean μ and variance σ^2, and let Φ be the cumulative distribution function of the standard normal distribution. Find the expected value of $Y = \Phi(X)$.

8. What is the law of large numbers?

9. What is the central limit theorem?

10. What is a martingale? How is it related to option pricing?

11. Explain the assumption $(dW_t)^2 = dt$ used in the informal derivation of Itô's Lemma.

12. If W_t is a Wiener process, find $E[W_t W_s]$.

13. If W_t is a Wiener process, what is $\mathrm{var}(W_t + W_s)$?

14. Let W_t be a Wiener process. Find

$$\int_0^t W_s \, dW_s \text{ and } E\left[\int_0^t W_s \, dW_s\right].$$

15. Find the distribution of the random variable

$$X = \int_0^1 W_t dW_t.$$

16. Let W_t be a Wiener process. Find the mean and the variance of

$$\int_0^t W_s^2 dW_s.$$

17. If W_t is a Wiener process, find the variance of

$$X = \int_0^1 \sqrt{t} e^{\frac{W_t^2}{8}} \, dW_t.$$

18. If W_t is a Wiener process, what is $E\left[e^{W_t}\right]$?

19. If W_t is a Wiener process, find the variance of

$$\int_0^t s \, dW_s.$$

20. Let W_t be a Wiener process, and let

$$X_t = \int_0^t W_\tau d\tau.$$

What is the distribution of X_t? Is X_t a martingale?[2]

21. What is an Itô process?

22. What is Itô's lemma?

23. If W_t is a Wiener process, is the process $X_t = W_t^2$ a martingale?

[2] A solution to this question was given in Chapter 2 using integration by parts; a different solution will be given herein.

24. If W_t is a Wiener process, is the process

$$N_t = W_t^3 - 3tW_t$$

a martingale?

25. What is Girsanov's theorem?

26. What is the martingale representation theorem, and how is it related to option pricing and hedging?

27. Solve $dY_t = Y_t\, dW_t$, where W_t is a Wiener process.

28. Solve the following SDEs:

    (i) $dY_t = \mu Y_t dt + \sigma Y_t dW_t$;

    (ii) $dX_t = \mu dt + (aX_t + b) dW_t$.

29. What is the Heston model?

4.7 Brainteasers.

1. A flea is going between two points which are 100 inches apart by jumping (always in the same direction) either one inch or two inches at a time. How many different paths can the flea travel by?

2. I have a bag containing three pancakes: one golden on both sides, one burnt on both sides, and one golden on one side and burnt on the other. You shake the bag, draw a pancake at random, look at one side, and notice that it is golden. What is the probability that the other side is golden?

3. Alice and Bob are playing heads and tails, Alice tosses $n+1$ coins, Bob tosses n coins. The coins are fair. What is the probability that Alice will have strictly more heads than Bob?

4. Alice is in a restaurant trying to decide between three desserts. How can she choose one of three desserts with equal probability with the help of a fair coin? What if the coin is biased and the bias is unknown?

5. What is the expected number of times you must flip a fair coin until it lands on head? What if the coin is biased and lands on head with probability p?

6. What is the expected number of coin tosses of a fair coin in order to get two heads in a row? What if

the coin is biased with 25% probability of getting heads?

7. A fair coin is tossed n times. What is the probability that no two consecutive heads appear?

8. You have two identical Fabergé eggs, either of which would break if dropped from the top of a building with 100 floors. Your task is to determine the highest floor from which an egg could be dropped without breaking. What is the minimum number of drops required to achieve this? You are allowed to break both eggs in the process.

9. An ant is in the corner of a $10 \times 10 \times 10$ room and wants to go to the opposite corner. What is the length of the shortest path the ant can take?

10. A $10 \times 10 \times 10$ cube is made of $1,000$ unit cubes. How many unit cubes can you see on the outside?

11. Fox Mulder is imprisoned by aliens in a large circular field surrounded by a fence. Outside the fence is a vicious alien that can run four times as fast as Mulder, but is constrained to stay near the fence. If Mulder can contrive to get to an unguarded point on the fence, he can quickly scale the fence and escape. Can he get to a point on the fence ahead of the alien?

12. At your subway station, you notice that of the two trains running in opposite directions which are supposed to arrive with the same frequency, the train

going in one direction comes first 80% of the time, while the train going in the opposite direction comes first only 20% of the time. What do you think could be happening?

13. You start off with one amoeba. Every minute, this amoeba can either die, do nothing, split into two amoebas, or split into three amoebas; all these scenarios being equally likely to happen. All further amoebas behave the same way. What is the probability that the amoebas eventually die off?

14. Given a set X with n elements, choose two subsets A and B at random. What is the probability of A being a subset of B?

15. Alice writes two distinct real numbers between 0 and 1 on two sheets of paper. Bob selects one of the sheets randomly to inspect it. He then has to declare whether the number he sees is the bigger or smaller of the two. Is there any way Bob can expect to be correct more than half the times Alice plays this game with him?

16. How many digits does the number 125^{100} have? You are not allowed to use values of $\log_{10} 2$ or $\log_{10} 5$.

17. For every subset of $\{1, 2, 3, \ldots, 2013\}$, arrange the numbers in the increasing order and take the sum with alternating signs. The resulting integer is called

the weight of the subset.[3] Find the sum of the weights of all the subsets of $\{1, 2, 3, \ldots, 2013\}$.

18. Alice and Bob alternately choose one number from one of the following nine numbers: 1/16, 1/8, 1/4, 1/2, 1, 2, 4, 8, 16, without replacement. Whoever gets three numbers that multiply to one wins the game. Alice starts first. What should her strategy be? Can she always win?

19. Mr. and Mrs. Jones invite four other couples over for a party. At the end of the party, Mr. Jones asks everyone else how many people they shook hands with, and finds that everyone gives a different answer. Of course, no one shook hands with his or her spouse and no one shook the same person's hand twice. How many people did Mrs. Jones shake hands with?

20. The New York Yankees and the San Francisco Giants are playing in the World Series (best of seven format). You would like to bet $100 on the Yankees winning the World Series, but you can only place bets on individual games, and every time at even odds. How much should you bet on the first game?

21. We have two red, two green and two yellow balls. For each color, one ball is heavy and the other is light. All heavy balls weigh the same. All light balls weigh the same. How many weighings on a scale are necessary to identify the three heavy balls?

[3] For example, the weight of the subset $\{3\}$ is 3. The weight of the subset $\{2, 5, 8\}$ is $2 - 5 + 8 = 5$.

22. There is a row of 10 rooms and a treasure in one of them. Each night, a ghost moves the treasure to an adjacent room. You are trying to find the treasure, but can only check one room per day. How do you find it?

23. How many comparisons do you need to find the maximum in a set of n distinct numbers? How many comparisons do you need to find both the maximum and minimum in a set of n distinct numbers?

24. Given a cube, you can jump from one vertex to a neighboring vertex with equal probability. Assume you start from a certain vertex (does not matter which one). What is the expected number of jumps to reach the opposite vertex?

25. Select numbers uniformly distributed between 0 and 1, one after the other, as long as they keep decreasing; i.e. stop selecting when you obtain a number that is greater than the previous one you selected.

26. To organize a charity event that costs \$100K, an organization raises funds. Independent of each other, one donor after another donates some amount of money that is exponentially distributed with a mean of \$20K. The process is stopped as soon as \$100K or more has been collected. Find the distribution, mean, and variance of the number of donors needed until at least \$100K has been collected.

27. Consider a random walk starting at 1 and with equal probability of moving to the left or to the right by one unit, and stopping either at 0 or at 3.

    (i) What is the expected number of steps to do so?

    (ii) What is the probability of the random walk ending at 3 rather than at 0?

28. A stick of length 1 drops and breaks at a random place uniformly distributed across the length. What is the expected length of the smaller part?

29. You are given a stick of unit length.

    (i) The stick drops and breaks at two places. What is the probability that the three pieces could form a triangle?

    (ii) The stick drops and breaks at one place. Then the larger piece is taken and dropped again, breaking at one place. What is the probability that the three pieces could form a triangle?

30. Why is a manhole cover round?

31. When is the first time after 12 o'clock that the hour and minute hands of a clock meet again?

32. Three light switches are in one room, and they turn three light bulbs in another. How do you figure out which switch turns on which bulb in one shot?

Chapter 5

Solutions

5.1 基础数学,微积分,微分方程.

题目1. 求 i^i 的值,其中 $i=\sqrt{-1}$.

参考答案: 已知 $e^{i\theta}=\cos\theta+i\sin\theta$. 则

$$i = \cos\frac{\pi}{2}+i\sin\frac{\pi}{2} = e^{i\frac{\pi}{2}},$$

并且 $i^2=-1$. 则有

$$i^i = \left(e^{i\frac{\pi}{2}}\right)^i = e^{(i\frac{\pi}{2})i} = e^{i^2\frac{\pi}{2}} = e^{-\frac{\pi}{2}},$$

□

题目2. 比较 π^e 与 e^π的大小.

参考答案: 需要证明 $\pi^e<e^\pi$. 不等式两边同时取对数, 则有

$$\begin{aligned}\pi^e<e^\pi &\iff \ln(\pi^e)<\ln(e^\pi) \iff e\ln(\pi)<\pi \\ &\iff \frac{\ln(\pi)}{\pi}<\frac{1}{e},\end{aligned} \tag{5.1}$$

也可以写成

$$\frac{\ln(\pi)}{\pi} < \frac{\ln(e)}{e}.$$

记函数 $f:(0,\infty)\to\mathbb{R}$ 为 $f(x)=\frac{\ln(x)}{x}$. 求导得

$$f'(x) = \frac{1-\ln(x)}{x^2}.$$

注意 $f'(x)=0$ 仅有一个解 $x=e$. 因为当 $0<x<e$ 时 $f'(x)>0$, 当 $x>e$ 时 $f'(x)<0$, 所以 $f(x)$ 在 $(0,e)$ 区间内递增同时在 (e,∞)区间内递减.

当 $x=e$时函数 $f(x)=\frac{\ln(x)}{x}$ 取到全局最大值, 那么当 $x>0$ 且 $x\neq e$时: $f(x)<f(e)=\frac{1}{e}$, 得到:

$$f(\pi) = \frac{\ln(\pi)}{\pi} < \frac{\ln(e)}{e} = \frac{1}{e},$$

上述不等式等价于 $\pi^e<e^\pi$; cf. (5.1). □

题目3. 证明

$$\frac{e^x+e^y}{2} \geq e^{\frac{x+y}{2}} \quad \forall\, x,y\in\mathbb{R}. \tag{5.2}$$

参考答案: 设 $e^x=a$, $e^y=b$. 且 $a,b>0$, 代入原式:

$$e^{\frac{x+y}{2}} = \sqrt{e^{x+y}} = \sqrt{e^x\cdot e^y} = \sqrt{ab}.$$

则(5.2) 可以写成

$$\begin{aligned}\frac{a+b}{2} \geq \sqrt{ab} &\iff a+b-2\sqrt{ab} \geq 0\\ &\iff \left(\sqrt{a}-\sqrt{b}\right)^2 \geq 0\end{aligned}$$

□

题目4. 解 $x^6=64$.

参考答案: $z^6 = 1$ 一共有六个解分别为:

$$\begin{aligned} z_k &= \exp\left(\frac{2k\pi i}{6}\right) = \exp\left(\frac{k\pi i}{3}\right) \\ &= \cos\left(\frac{k\pi}{3}\right) + i\sin\left(\frac{k\pi}{3}\right), \end{aligned} \tag{5.3}$$

其中$k = 0:5$. 因为$\sqrt[6]{64} = 2$, 根据(5.3),原方程的解为:

$$x_k = 2\cos\left(\frac{k\pi}{3}\right) + 2i\sin\left(\frac{k\pi}{3}\right), \quad \forall\, k = 0:5. \quad \square$$

题目5. 求 x^x 的导数.

参考答案: 对x^x做如下变换:

$$x^x = e^{\ln(x^x)} = e^{x\ln(x)}. \tag{5.4}$$

使用链式求导法则(5.4), 我们可以得到

$$\begin{aligned} (x^x)' &= \left(e^{x\ln(x)}\right)' = e^{x\ln(x)}\,(x\ln(x))' \\ &= x^x(\ln(x) + 1). \quad \square \end{aligned}$$

题目6. 计算

$$\sqrt{2 + \sqrt{2 + \sqrt{2 + \ldots}}} \tag{5.5}$$

参考答案: 假设式 (5.5) 存在极限l, 则 $l = \sqrt{2 + l}$. 整理后得:

$$l^2 - l - 2 = (l-2)(l+1) = 0.$$

又因为 $l > 0$,我们得到 $l = 2$,即

$$\sqrt{2 + \sqrt{2 + \sqrt{2 + \ldots}}} = 2. \tag{5.6}$$

要证明等式(5.6) 成立,等价于证明序列 $(x_n)_{n\geq 0}$收敛,其中$x_0 = \sqrt{2}$,

$$x_{n+1} = \sqrt{2 + x_n}, \ \forall\, n \geq 0,$$

我们可以由数学归纳法证明$(x_n)_{n\geq 0}$存在上界2. 因为 $x_0 = \sqrt{2} < 2$, 如果假设 $x_n < 2$, 则

$$x_{n+1} = \sqrt{2 + x_n} < \sqrt{4} = 2.$$

此外,序列$(x_n)_{n\geq 0}$是递增的,因为

$$\begin{aligned} x_n < x_{n+1} &\iff x_n < \sqrt{2 + x_n} \\ &\iff x_n^2 - x_n - 2 < 0 \\ &\iff (x_n - 2)(x_n + 1) < 0, \end{aligned}$$

上式一定成立,因为$x_n > 0$,并且我们已经证明对任意$n > 0$, $x_n < 2$.

因此,序列 $(x_n)_{n\geq 0}$ 有上界并且单调递增,因此一定收敛. 我们由此证明了式(5.6). □

题目7. 找到使下式成立的x

$$x^{x^{x^{\cdot^{\cdot^{\cdot}}}}} = 2. \tag{5.7}$$

参考答案: 如果存在使得 (5.7) 成立的x, 则

$$x^{x^{x^{x^{\cdot^{\cdot^{\cdot}}}}}} = x^2 = 2,$$

有$x = \sqrt{2}$. 下面我们要证明 $x = \sqrt{2}$ 是 (5.7) 的解.需要证明下列式子成立.

$$\sqrt{2}^{\sqrt{2}^{\sqrt{2}^{\cdot^{\cdot^{\cdot}}}}} = 2. \tag{5.8}$$

定义一个序列$(x_n)_{n\geq 0}$, 其中 $x_0 = \sqrt{2}$, 且满足如下递推关系

$$x_{n+1} = \left(\sqrt{2}\right)^{x_n} = 2^{x_n/2}, \ \forall\, n \geq 0. \tag{5.9}$$

因为$x_0 = \sqrt{2} < \sqrt{2}^{\sqrt{2}} = x_1$,并且

$$x_n = 2^{x_{n-1}/2} \ < \ 2^{x_n/2} = x_{n+1}.$$

则$(x_n)_{n\geq 0}$ 是一个递增序列.

根据数学归纳法, 我们可以得到 $(x_n)_{n\geq 0}$ 存在上界 2. 因为$x_0 = \sqrt{2} < 2$, 如果假设$x_n < 2$,有

$$x_{n+1} = 2^{x_n/2} \ < \ 2.$$

由于上述序列为递增序列则 $(x_n)_{n\geq 0}$ 收敛.

设 $l = \lim_{n\to\infty} x_n$. 由式(5.9)有 $l = 2^{l/2}$,等价于

$$l^{1/l} \ = \ 2^{1/2}. \tag{5.10}$$

定义函数 $f : (0, \infty) \to (0, \infty)$

$$\begin{aligned} f(t) \ &= \ t^{1/t} \ = \ \exp\left(\ln(t^{1/t})\right) \\ &= \ \exp\left(\frac{\ln(t)}{t}\right) \end{aligned}$$

求导得

$$f'(t) \ = \ \frac{1 - \ln(t)}{t^2} \exp\left(\frac{\ln(t)}{t}\right)$$

当$t < e$ 时 $f'(t) > 0$;当 $t > e$ 时 $f'(t) < 0$,即当$t < e$ 时 $f(t)$ 递增,当 $t > e$时 $f(t)$ 递减.

因此,式(5.10)有两个解. 2为其中一个解,另一个解则大于e. 因为对任意$n \geq 0$, $x_n < 2$. 则 $l = \lim_{n\to\infty} x_n = 2$时,使式 (5.8)成立. □

题目8. 下列序列是否收敛?

$$\sum_{k=1}^{\infty} \frac{1}{k}; \quad \sum_{k=1}^{\infty} \frac{1}{k^2}; \quad \sum_{k=2}^{\infty} \frac{1}{k\ln(k)}.$$

参考答案: 我们将证明

$$\sum_{k=1}^{\infty} \frac{1}{k^2} \quad 收敛;$$

$$\sum_{k=1}^{\infty} \frac{1}{k} \quad 和 \quad \sum_{k=2}^{\infty} \frac{1}{k\ln(k)} \quad 不收敛.$$

因为 $\sum_{k=1}^{\infty} \frac{1}{k^2}$ 的所有项大于零, 所以只需要证明 $\sum_{k=1}^{n} \frac{1}{k^2}$ 是一致有界的, 则它一定收敛.证明如下:

$$\begin{aligned}
\sum_{k=1}^{n} \frac{1}{k^2} &= 1 + \sum_{k=2}^{n} \frac{1}{k^2} \\
&\leq 1 + \sum_{k=2}^{n} \frac{1}{k(k-1)} \\
&= 1 + \sum_{k=2}^{n} \left(\frac{1}{k-1} - \frac{1}{k} \right) \\
&= 1 + \left(1 - \frac{1}{n} \right) \\
&< 2, \quad \forall\ n \geq 2.
\end{aligned}$$

为了证明 $\sum_{k=1}^{\infty} \frac{1}{k}$ 不收敛, 先证明如下不等式

$$\sum_{k=1}^{n} \frac{1}{k} > \ln(n) + \frac{1}{n}, \ \forall\, n \geq 1. \tag{5.11}$$

因为 $\frac{1}{x}$ 是单调递减函数,有

$$\frac{1}{x} < \frac{1}{k}, \ \forall\, k < x < k+1.$$

并且

$$\begin{aligned}
\int_1^n \frac{1}{x}\,dx &= \sum_{k=1}^{n-1}\int_k^{k+1}\frac{1}{x}\,dx \\
&< \sum_{k=1}^{n-1}\int_k^{k+1}\frac{1}{k}\,dx \\
&= \sum_{k=1}^{n-1}\frac{1}{k} \\
&= -\frac{1}{n}+\sum_{k=1}^{n}\frac{1}{k}. \qquad (5.12)
\end{aligned}$$

通过式 (5.12),有

$$\begin{aligned}
\sum_{k=1}^{n}\frac{1}{k} &> \int_1^n \frac{1}{x}\,dx+\frac{1}{n} \\
&= \ln(n)+\frac{1}{n},\ \forall\, n\geq 1,
\end{aligned}$$

得证(5.11).

同理,注意到

$$\frac{1}{x\ln(x)} < \frac{1}{k\ln(k)},\ \forall\, k<x<k+1,$$

因此

$$\begin{aligned}
\int_2^{n+1}\frac{1}{x\ln(x)}\,dx &= \sum_{k=2}^{n}\int_k^{k+1}\frac{1}{x\ln(x)}dx \\
&< \sum_{k=2}^{n}\int_k^{k+1}\frac{1}{k\ln(k)}dx \\
&= \sum_{k=2}^{n}\frac{1}{k\ln(k)}. \qquad (5.13)
\end{aligned}$$

又因为

$$\int_{2}^{n+1} \frac{1}{x\ln(x)}\,dx \;=\; \ln(\ln(n+1)) - \ln(\ln(2)),$$

根据 (5.13) 得到

$$\sum_{k=2}^{n} \frac{1}{k\ln(k)} \;>\; \ln(\ln(n+1)) - \ln(\ln(2)),$$

则序列 $\sum_{k=2}^{\infty} \frac{1}{k\ln(k)}$ 不收敛.

注:虽然题目没有要求,但我们可以算出

$$\sum_{k=1}^{\infty} \frac{1}{k^2} \;=\; \frac{\pi^2}{6},$$

$$\lim_{n\to\infty} \left(\sum_{k=1}^{\infty} \frac{1}{k} \;-\; \ln(n) \right) \;=\; \gamma,$$

其中 $\gamma \approx 0.57721$ 是欧拉常数. □

题目9. 计算

$$\int \frac{1}{1+x^2}\,dx.$$

参考答案: 对上式做如下变量替换 $x = \tan(z)$, 则 $dx = \frac{1}{\cos^2(z)}dz$

$$\begin{aligned}
\int \frac{1}{1+x^2}\,dx &= \int \frac{1}{(1+\tan^2(z))\cos^2(z)}\,dz \\
&= \int \frac{1}{\cos^2(z)+\sin^2(z)}\,dz \\
&= \int 1\,dz \\
&= z \,+\, C,
\end{aligned}$$

其中 C 是常数,

因为 $x = \tan(z)$, 则 $z = \arctan(x)$. 将其代入原式后得到结果.

$$\int \frac{1}{1+x^2}\,dx = \arctan(x) + C. \quad \square$$

题目10. 计算

$$\int x\ln(x)dx \quad \text{and} \quad \int xe^x dx.$$

参考答案: 使用分部积分:

$$\begin{aligned}
\int x\ln(x)\,dx &= \frac{x^2}{2}\ln(x) - \int \frac{x^2}{2}\cdot\frac{1}{x}\,dx \\
&= \frac{x^2}{2}\ln(x) - \frac{1}{2}\int x\,dx \\
&= \frac{x^2\ln(x)}{2} - \frac{x^2}{4} + C; \\
\int xe^x\,dx &= xe^x - \int 1\cdot e^x\,dx \\
&= xe^x - e^x + C. \quad \square
\end{aligned}$$

题目11. 计算

$$\int x^n\ln(x)\,dx.$$

参考答案: 当 $n \neq -1$时, 使用分部积分:

$$\begin{aligned}
\int x^n\ln(x)\,dx &= \frac{x^{n+1}}{n+1}\ln(x) - \int \frac{x^{n+1}}{n+1}\cdot\frac{1}{x}\,dx \\
&= \frac{x^{n+1}\ln(x)}{n+1} - \frac{1}{n+1}\int x^n\,dx \\
&= \frac{x^{n+1}\ln(x)}{n+1} - \frac{x^{n+1}}{(n+1)^2} + C.
\end{aligned}$$

当 $n=-1$时

$$\left(\frac{(\ln(x))^2}{2}\right)' = \frac{1}{2}\cdot 2\ln(x)\cdot(\ln(x))' = \frac{\ln(x)}{x}.$$

$$\int \frac{\ln(x)}{x}\,dx = \frac{(\ln(x))^2}{2} + C$$

这里 C 均为常数. □

题目12. 计算

$$\int (\ln(x))^n\,dx.$$

参考答案: 当 $n\geq 0$时记

$$f_n(x) = \int (\ln(x))^n\,dx.$$

使用分部积分得到

$$\begin{aligned}
&\int (\ln(x))^n\,dx \\
= \quad & x(\ln(x))^n - \int x\left((\ln(x))^n\right)'\,dx \\
= \quad & x(\ln(x))^n - \int x\cdot n(\ln(x))^{n-1}\cdot(\ln(x))'\,dx \\
= \quad & x(\ln(x))^n - \int x\cdot n(\ln(x))^{n-1}\cdot\frac{1}{x}\,dx \\
= \quad & x(\ln(x))^n - n\int (\ln(x))^{n-1}\,dx,
\end{aligned}$$

所以

$$f_n(x) = x(\ln(x))^n - nf_{n-1}(x),\ \ \forall\, n\geq 1. \qquad (5.14)$$

又当n取0时有

$$f_0(x) = \int 1dx = x + C.$$

根据 (5.14)的递推关系, 我们可以求出当任意 $n \geq 0$时 $f_n(x)$的值. 例如:

$$\begin{aligned} f_1(x) &= x\ln(x) - f_0(x) = x(\ln(x)-1)+C; \\ f_2(x) &= x(\ln(x))^2 - 2f_1(x) \\ &= x\left((\ln(x))^2 - 2\ln(x) + 2\right) + C. \end{aligned}$$

根据归纳法, 我们得到如下通项形式

$$\int (\ln(x))^n \, dx = x\sum_{k=0}^{n} \frac{(-1)^{n-k} n!}{k!} (\ln(x))^k + C.$$

□

题目13. 求解常微分方程

$$y'' - 4y' + 4y = 1. \tag{5.15}$$

参考答案: 式 (5.15) 是系数为常数的二阶线性非齐次方程. 它对应的齐次方程为

$$y'' - 4y' + 4y = 0, \tag{5.16}$$

特征方程为 $z^2 - 4z + 4 = 0$有一个二重根 $z_1 = z_2 = 2$. 因此齐次方程 (5.16) 的通解为

$$y(x) = c_1 e^{2x} + c_2 x e^{2x}, \tag{5.17}$$

其中 c_1, c_2 为常数.

由于$y_0(x) = \frac{1}{4}$ 是非齐次方程 (5.15)的一个特解, 因此(5.15)的通解为

$$y(x) = c_1 e^{2x} + c_2 x e^{2x} + \frac{1}{4}. \quad \Box$$

题目14. 找到$f(x)$使得

$$f'(x) = f(x)(1 - f(x)). \tag{5.18}$$

参考答案: 式(5.18) 可以分离变量, 写成如下形式

$$\frac{y'}{y(1-y)} = 1, \tag{5.19}$$

其中$y = f(x)$. 对 (5.19) 求x的积分可以得到

$$\int \frac{y'}{y(1-y)}\,dx = \int 1\,dx = x + C_1, \tag{5.20}$$

其中$C_1 \in \mathbb{R}$ 为常数.

又因为$dy = y'dx$, 则

$$\begin{aligned}
\int \frac{y'}{y(1-y)}\,dx &= \int \frac{1}{y(1-y)}\,dy \\
&= \int \left(\frac{1}{y} + \frac{1}{1-y}\right)\,dy \\
&= \int \frac{1}{y}\,dy + \int \frac{1}{1-y}\,dy \\
&= \ln(|y|) - \ln(|1-y|) \\
&= \ln\left|\frac{y}{1-y}\right|.
\end{aligned} \tag{5.21}$$

从式 (5.20) 和式 (5.21), 我们得到

$$\ln\left|\frac{y}{1-y}\right| = x + C_1,$$

从而

$$\left|\frac{y}{1-y}\right| = e^{x+C_1} = C_2 e^x,$$

其中 $C_2 = e^{C_1} > 0$ 为大于0的常数.

从而得到两个通解分别为 $\frac{y}{1-y} = C_2 e^x$, $\frac{y}{1-y} = -C_2 e^x$, 又可以写成

$$\frac{y}{1-y} = Ce^x, \tag{5.22}$$

其中 C 为常数.

注意到 $(D_{\sigma_X})^{-1} = \mathrm{diag}\left(\frac{1}{\sigma_i}\right)_{i=1:n}$. 令 $v \in \mathbb{R}^n$为n维随机变量,设

$$w = (D_{\sigma_X})^{-1} v.$$

则

$$w^t = v^t \left((D_{\sigma_X})^{-1}\right)^t = v^t (D_{\sigma_X})^{-1}, \tag{5.34}$$

因为对角矩阵是对称矩阵,所以 $(D_{\sigma_X})^{-1}$ 是对称矩阵, 则 $\left((D_{\sigma_X})^{-1}\right)^t = (D_{\sigma_X})^{-1}$.

通过 (5.33), (5.34), 和 (5.30), 我们有

$$\begin{aligned}
w^t \Sigma_X w &= w^t (D_{\sigma_X} \Omega_X D_{\sigma_X}) w \\
&= v^t (D_{\sigma_X})^{-1} (D_{\sigma_X} \Omega_X D_{\sigma_X}) (D_{\sigma_X})^{-1} v \\
&= v^t \Omega_X v \\
&\geq 0.
\end{aligned}$$

因此,对于任意$v \in \mathbb{R}^n$均有 $v^t \Omega_X v \geq 0$. 那么我们得证: Ω_X 是对称半正定矩阵. □

题目2. 已知三个随机变量的协方差矩阵如下, 求它们的相关系数矩阵.

$$\Sigma_X = \begin{pmatrix} 1 & 0.36 & -1.44 \\ 0.36 & 4 & 0.80 \\ -1.44 & 0.80 & 9 \end{pmatrix}. \tag{5.35}$$

参考答案: 设 Ω_X 是三个随机变量的相关系数矩阵,则

$$\Sigma_X = \begin{pmatrix} \sigma_1 & 0 & 0 \\ 0 & \sigma_2 & 0 \\ 0 & 0 & \sigma_3 \end{pmatrix} \Omega_X \begin{pmatrix} \sigma_1 & 0 & 0 \\ 0 & \sigma_2 & 0 \\ 0 & 0 & \sigma_3 \end{pmatrix},$$

其中 σ_i, $i = 1:3$, 分别是三个随机变量的标准差(方差开平方).

$$\Omega_X = \begin{pmatrix} \frac{1}{\sigma_1} & 0 & 0 \\ 0 & \frac{1}{\sigma_2} & 0 \\ 0 & 0 & \frac{1}{\sigma_3} \end{pmatrix} \Sigma_X \begin{pmatrix} \frac{1}{\sigma_1} & 0 & 0 \\ 0 & \frac{1}{\sigma_2} & 0 \\ 0 & 0 & \frac{1}{\sigma_3} \end{pmatrix}.$$

可以通过如下的等式从协方差矩阵Σ_X 中得到每一个随机变量的标准差 (5.35)

$$\begin{aligned}\sigma_1 &= \sqrt{\Sigma_X(1,1)} = 1;\\ \sigma_2 &= \sqrt{\Sigma_X(2,2)} = 2;\\ \sigma_3 &= \sqrt{\Sigma_X(2,2)} = 3,\end{aligned}$$

从而协方差矩阵为

$$\begin{aligned}\Omega_X &= \begin{pmatrix} 1 & 0 & 0\\ 0 & \frac{1}{2} & 0\\ 0 & 0 & \frac{1}{3}\end{pmatrix} \Sigma_X \begin{pmatrix} 1 & 0 & 0\\ 0 & \frac{1}{2} & 0\\ 0 & 0 & \frac{1}{3}\end{pmatrix}\\ &= \begin{pmatrix} 1 & 0.18 & -0.48\\ 0.18 & 1 & 0.1333\\ -0.48 & 0.1333 & 1\end{pmatrix}. \quad \square\end{aligned}$$

题目3. 假设 $n \times n$ 相关系数矩阵的所有非主对角元素均为ρ.求ρ的取值范围.

参考答案: 已知相关系数矩阵的对角元素都为 1.且相关系数矩阵必定是对称半正定矩阵,其所有特征值非负.根据这个性质可以得到使 Ω 为相关系数矩阵的条件.我们介绍两种计算Ω的特征值的方法.首先不妨设矩阵形式为:

$$\Omega = \begin{pmatrix} 1 & \rho & \dots & \rho\\ \rho & \ddots & \ddots & \vdots\\ \vdots & \ddots & \ddots & \rho\\ \rho & \dots & \rho & 1\end{pmatrix}.$$

解法 1:

$$\Omega = (1-\rho)I + \begin{pmatrix} \rho & \rho & \cdots & \rho \\ \rho & \ddots & \ddots & \vdots \\ \vdots & \ddots & \ddots & \rho \\ \rho & \cdots & \rho & \rho \end{pmatrix}$$
$$= (1-\rho)I + \rho M,$$

M 为 $n \times n$的方阵,其所有元素均为1,I 是 $n \times n$ 的单位矩阵.

设 λ 和 $v = (v_i)_{i=1:n}$ 分别为矩阵M的特征值与特征向量. 则有$Mv = \lambda v$, 且 $v \neq 0$.

我们可以把$Mv = \lambda v$写成如下的形式：

$$\begin{cases} v_1 + v_2 + \cdots + v_n = \lambda v_1; \\ v_1 + v_2 + \cdots + v_n = \lambda v_2; \\ \vdots \qquad\qquad \vdots \\ v_1 + v_2 + \cdots + v_n = \lambda v_n, \end{cases}$$

解得:

$$\lambda v_1 = \lambda v_2 = \ldots = \lambda v_n$$

因此M有两个不同的特征值分别为$\lambda = 0$和$\lambda = n$. 并且当$\lambda = n$时有$v_1 = v_2 = \cdots = v_n$.

将上述结果代入特征向量的定义$Mv = \lambda v$可以得到:

$$\Omega v = (1-\rho)v + \rho M v = (1-\rho)v + \rho\lambda v$$
$$= (1-\rho+\rho\lambda)v.$$

则$\mu = 1-\rho+\rho\lambda$为Ω的特征值. 且 v 为Ω的特征向量. 因为M 的特征值为$\lambda = 0$ 和 $\lambda = n$, 则Ω 的特征值[1] 分别为当 $\lambda = 0$时$\mu = 1-\rho$, 和当 $\lambda = n$时$\mu = (1-\rho) + n\rho = 1 + (n-1)\rho$.

[1]其中 $1-\rho$ 为 $n-1$重根, 而 $1+(n-1)\rho$ 没有重根. 解法 2 中有详细说明.

综上所述使 Ω 是相关系数矩阵的充要条件是它的所有特征值非负, 即要满足

$$0 \leq 1 + (n-1)\rho \quad \text{和} \quad 0 \leq 1 - \rho,$$

等价于

$$-\frac{1}{n-1} \leq \rho \leq 1. \tag{5.36}$$

解法 2: 已知相关系数矩阵可以写成:

$$\begin{aligned}
\Omega &= (1-\rho)I + \rho \begin{pmatrix} 1 & 1 & \dots & 1 \\ 1 & \ddots & \ddots & \vdots \\ \vdots & \ddots & \ddots & 1 \\ 1 & \dots & 1 & 1 \end{pmatrix} \\
&= (1-\rho)I + \rho \begin{pmatrix} 1 \\ \vdots \\ 1 \end{pmatrix} (1 \dots 1) \\
&= (1-\rho)I + \rho w w^t \\
&= (1-\rho)I + \rho A,
\end{aligned}$$

其中 I 为 $n \times n$ 的单位矩阵; A 为 $n \times n$ 的方阵,其元素均为1. 我们观察到矩阵A可以写成 $A = ww^t$的形式, 其中w为所有元素为1, $n \times 1$ 的列向量.

注意到如果 $n \times n$ 的矩阵可以写成 uu^t形式,其中$u = (u_i)_{i=1:n}$ 是 $n \times 1$ 的列向量, 那么该矩阵有两个不同的特征值,其中一个为 $\sum_{i=1}^n u_i^2$ 这个根没有重根.另外一个根为 0, 为$n-1$重根.

因此我们得到矩阵A的特征值分别为: $\lambda = \sum_{i=1}^n w_i^2 = \sum_{i=1}^n 1 = n$ 这个根没有重根; $\lambda = 0$ 这个根为$n-1$重根.

因为 λ 和 v 分别为A的特征值和特征向量. 根据定义有$Av = \lambda v$,所以

$$\begin{aligned}
\Omega v &= (1-\rho)v + \rho A v = (1-\rho)v + \rho\lambda v \\
&= (1 - \rho + \rho\lambda)v.
\end{aligned}$$

因此 $1-\rho+\rho\lambda$ 和 v 分别为Ω的特征值和特征向量. 我们得到Ω的特征值为:

- $(1-\rho)+n\rho = 1+(n-1)\rho$ 为 1重根;
- $1-\rho$ 为 $n-1$重根.

和前面一样,使 Ω 是相关系数矩阵的充要条件是

$$0 \le 1+(n-1)\rho \quad 和 \quad 0 \le 1-\rho,$$

即

$$-\frac{1}{n-1} \le \rho \le 1,$$

与 (5.36)的结果相同. □

题目 4. $n \times n$的实矩阵有几个特征值和特征向量?

参考答案: $n \times n$ 实矩阵有 n 个特征值(算上重数);特征值可能为复数. $n \times n$ 矩阵最多有 n 个特征向量.

设 A为 $n \times n$ 矩阵, λ 为A的特征值,并且它对应的特征向量为 $v \neq 0$. 不妨设 $P_A(x) = \det(xI_n - A)$ 为A的特征多项式,其中 I_n 为 $n \times n$的单位矩阵. 则:

$$\begin{aligned} Av = \lambda v,\ v \neq 0 \quad &\Longleftrightarrow \quad (\lambda I_n - A)v = 0,\ v \neq 0 \\ &\Longleftrightarrow \quad \lambda I_n - A \text{ 为奇异矩阵} \\ &\Longleftrightarrow \quad \det(\lambda I_n - A) = 0 \\ &\Longleftrightarrow \quad P_A(\lambda) = 0. \end{aligned}$$

根据线性代数定理,λ 是 A的特征值的充要条件是 λ 是特征多项式$P_A(x)$的根. 又因为 $P_A(x)$ 是 n阶多项式,并且根据代数的基本定理,n阶多项式必有n个根(把重根和虚数根计算在内).所以任意 $n \times n$ 实数矩阵有n 个特征值.

一个 m重的特征值,会有至少一个并且至多m个线性不相关的特征向量[2]. □

[2]例如有这样的矩阵: $\begin{pmatrix} 2 & 1 & 0 \\ 0 & 2 & 1 \\ 0 & 0 & 2 \end{pmatrix}$ 它有一个特征值等于2而这个特征值有3 重但是有且仅有一个特征向量. $\begin{pmatrix} 1 \\ 0 \\ 0 \end{pmatrix}$.

题目5. 已知

$$A = \begin{pmatrix} 2 & -2 \\ -2 & 5 \end{pmatrix}.$$

(i) 找到 2×2 的矩阵 M 使其满足 $M^2 = A$;

(ii) 找到 2×2 的矩阵 M 使其满足 $A = MM^t$.

参考答案: (i) 对于任意的对称矩阵,同学们应当首先想到如下的性质:

$$A = Q\Lambda Q^t, \tag{5.37}$$

其中Λ 是对角矩阵,对角元素为 A 的特征值,Q 为正交矩阵(orthogonal)由 A 的单位特征向量组成(向量模为1的特征向量).可以写成如下形式:

$$\Lambda = \begin{pmatrix} \lambda_1 & 0 \\ 0 & \lambda_2 \end{pmatrix}; \quad Q = (v_1 \ \ v_2), \tag{5.38}$$

其中 $Av_1 = \lambda_1 v_1$, $Av_2 = \lambda_2 v_2$, 有 $||v_1|| = ||v_2|| = 1$.

如果矩阵A 的特征值都非负, 即 $\lambda_1 \geq 0$ 并且 $\lambda_2 \geq 0$, 则本题的答案为(可以想象成对矩阵进行“开根号”运算)

$$M = Q\Lambda^{1/2}Q^t \tag{5.39}$$

其中

$$\Lambda^{1/2} = \begin{pmatrix} \sqrt{\lambda_1} & 0 \\ 0 & \sqrt{\lambda_2} \end{pmatrix} \tag{5.40}$$

这样的 M 满足 $M^2 = A$:

$$\begin{aligned} M^2 &= \left(Q\Lambda^{1/2}Q^t\right)\left(Q\Lambda^{1/2}Q^t\right) \\ &= Q\Lambda^{1/2}(Q^tQ)\Lambda^{1/2}Q^t \\ &= Q\Lambda^{1/2}\Lambda^{1/2}Q^t \\ &= Q\Lambda Q^t \\ &= A, \end{aligned}$$

由于 Q 是正交矩阵所以 $Q^tQ = I$, 又根据 (5.40), 得到 $\Lambda^{1/2}\Lambda^{1/2} = \Lambda$.

接下来需要求出 A的特征值和特征向量.根据上一题的方法, 矩阵的特征值为其特征多项式 $P_A(x)$ 的解,展开为[3]:

$$\begin{aligned} P_A(x) &= \det(xI - A) = \det\begin{pmatrix} x-2 & 2 \\ 2 & x-5 \end{pmatrix} \\ &= (x-2)(x-5) - 4 = x^2 - 7x + 6 \\ &= (x-1)(x-6). \end{aligned}$$

则$P_A(x)$ 的根为分别为 $\lambda_1 = 1$, $\lambda_2 = 6$, 同样这两个根也是矩阵 A 的特征值.

它们对应的单位特征向量为

$$v_1 = \begin{pmatrix} \frac{2}{\sqrt{5}} \\ \frac{1}{\sqrt{5}} \end{pmatrix} \quad 和 \quad v_2 = \begin{pmatrix} \frac{1}{\sqrt{5}} \\ -\frac{2}{\sqrt{5}} \end{pmatrix}.$$

下面我们举一个例子让大家熟悉一下如何在已知特征值的情况下求出特征向量.对于 $\lambda_2 = 6$所对应的特征值,设$v_2 = \begin{pmatrix} a \\ b \end{pmatrix} \neq 0$ 满足定义,则 $Av = 6v$, 代入上述公式后得:

$$\left\{ \begin{array}{rcl} 2a - 2b &=& 6a \\ -2a + 5b &=& 6b \end{array} \right. \Longleftrightarrow \left\{ \begin{array}{rcl} b &=& -2a \\ b &=& -2a \end{array} \right.$$

解出方程组后得到 $\lambda_2 = 6$ 对应的特征向量有如下形式

$$v_2 = \begin{pmatrix} a \\ -2a \end{pmatrix} = a\begin{pmatrix} 1 \\ -2 \end{pmatrix}.$$

[3]矩阵 A的特征多项式可以使用如下公式求得:

$$P_A(x) = x^2 - \mathrm{tr}(A)x + \det(A) = x^2 - 7x + 6,$$

其中 $\mathrm{tr}(A) = 2 + 5 = 7$, $\det(A) = 2 \cdot 5 - (-2) \cdot (-2) = 6$.

令 $a = \frac{1}{\sqrt{5}}$时得到了模为1的单位特征向量.所以$\lambda_2 = 6$ 所对应的单位特征向量是

$$v_2 = \begin{pmatrix} \frac{1}{\sqrt{5}} \\ -\frac{2}{\sqrt{5}} \end{pmatrix}.$$

代入 (5.39) 和 (5.40) 目标M 矩阵终于被找到了,

$$\begin{aligned} M &= \begin{pmatrix} \frac{2}{\sqrt{5}} & \frac{1}{\sqrt{5}} \\ \frac{1}{\sqrt{5}} & -\frac{2}{\sqrt{5}} \end{pmatrix} \begin{pmatrix} \sqrt{1} & 0 \\ 0 & \sqrt{6} \end{pmatrix} \begin{pmatrix} \frac{2}{\sqrt{5}} & \frac{1}{\sqrt{5}} \\ \frac{1}{\sqrt{5}} & -\frac{2}{\sqrt{5}} \end{pmatrix} \\ &= \frac{1}{5} \begin{pmatrix} 2 & 1 \\ 1 & -2 \end{pmatrix} \begin{pmatrix} 1 & 0 \\ 0 & \sqrt{6} \end{pmatrix} \begin{pmatrix} 2 & 1 \\ 1 & -2 \end{pmatrix} \\ &= \frac{1}{5} \begin{pmatrix} 4+\sqrt{6} & 2-2\sqrt{6} \\ 2-2\sqrt{6} & 1+4\sqrt{6} \end{pmatrix} \end{aligned}$$

它满足$M^2 = A$.

(ii) 通过方法一我们知道A 的两个特征值分别为 1 和 6它们都是正数. 所以 A是一个对称半正定矩阵,对于对称半正定矩阵我们可以使用楚列斯基分解算法(cholesky decomposition). 设矩阵U 是矩阵 A的楚列斯基因子(cholesky factor) 根据定义有$A = U^tU$. 为了找到满足$A = MM^t$的矩阵M,只需要求出楚列斯基因子(cholesky factor). 设楚列斯基因子(cholesky factor)为:

$$U = \begin{pmatrix} U(1,1) & U(1,2) \\ 0 & U(2,2) \end{pmatrix}$$

矩阵$A = U^tU$ 可以写成

$$\begin{pmatrix} A(1,1) & A(1,2) \\ A(2,1) & A(2,2) \end{pmatrix} \tag{5.41}$$

$$= \begin{pmatrix} U(1,1) & 0 \\ U(1,2) & U(2,2) \end{pmatrix} \begin{pmatrix} U(1,1) & U(1,2) \\ 0 & U(2,2) \end{pmatrix}.$$

将(5.41)展开后得到:

$$\begin{aligned} U(1,1) &= \sqrt{A(1,1)} = \sqrt{2}; \\ U(1,2) &= \frac{A(1,2)}{U(1,1)} = \frac{-2}{\sqrt{2}} = -\sqrt{2}; \\ U(2,2) &= \sqrt{A(2,2) - (U(1,2))^2} \\ &= \sqrt{5 - (-\sqrt{2})^2} \\ &= \sqrt{3} \end{aligned}$$

因此矩阵A 的楚列斯基因子为

$$U = \begin{pmatrix} \sqrt{2} & -\sqrt{2} \\ 0 & \sqrt{3} \end{pmatrix},$$

根据 $M = U^t$ 则有

$$M = \begin{pmatrix} \sqrt{2} & 0 \\ -\sqrt{2} & \sqrt{3} \end{pmatrix},$$

满足 $A = MM^t$. □

题目6. 已知 2×2 的矩阵 A 的特征值为 2 和 -3, 对应的特征向量分别为 $\begin{pmatrix} 1 \\ 2 \end{pmatrix}$ 和 $\begin{pmatrix} -1 \\ 3 \end{pmatrix}$. 设 $v = \begin{pmatrix} 3 \\ 1 \end{pmatrix}$, 求 Av.

参考答案: 已知 $\lambda_1 = 2$, $v_1 = \begin{pmatrix} 1 \\ 2 \end{pmatrix}$,和 $\lambda_2 = -3$, $v_2 = \begin{pmatrix} -1 \\ 3 \end{pmatrix}$. 根据特征值与特征向量的性质,两个不同特征值所对应的特征向量不会线性相关.并且在二维空间内的任何一个向量都可以用任意两个线性无关的向量线性表示.即存在两个常数, $c_1, c_2 \in \mathbb{R}$ 使得 $v = c_1v_1 + c_2v_2$, 代入已知条件后得到:

$$\begin{cases} 3 = c_1 - c_2 \\ 1 = 2c_1 + 3c_2 \end{cases}$$

解出方程组后得到 $c_1 = 2$ 和 $c_2 = -1$. 则

$$v = 2v_1 - v_2. \tag{5.42}$$

又根据特征值与特征向量的性质有: $Av_1 = \lambda_1 v_1 = 2v_1$ $Av_2 = \lambda_2 v_2 = -3v_2$, 那么(5.42) 可以写成

$$\begin{aligned} Av &= 2Av_1 - Av_2 = 2(2v_1) - (-3v_2) \\ &= 4v_1 + 3v_2 \\ &= \begin{pmatrix} 1 \\ 17 \end{pmatrix}. \quad \square \end{aligned}$$

题目7. 已知矩阵A和矩阵B是相同大小的方阵. 证明矩阵AB的迹和矩阵BA的迹相同.

参考答案: 回忆如数家珍的大学线性代数知识,对于大小相同的矩阵 A 和 B ,它们以不同顺序乘起来所得到的两个矩阵 AB 和 BA 有相同的特征方程式(特征多项式). 即:

$$\begin{aligned} P_{AB}(x) &= \det(xI - AB) = \det(xI - BA) \\ &= P_{BA}(x), \ \forall x \in \mathbb{R}, \end{aligned} \tag{5.43}$$

其中 I 为和方阵A 、B大小相同的单位矩阵.

并且$n \times n$ 矩阵M的特征多项式$P_M(x)$ 和它自己的迹$\mathrm{tr}(M)$有如下关系:

$$\begin{aligned} & P_M(x) \\ = \ & \det(xI - M) \\ = \ & x^n - \mathrm{tr}(M)x^{n-1} + \cdots + (-1)^n \det(M). \end{aligned} \tag{5.44}$$

且 $P_{AB}(x) = P_{BA}(x)$则根据 (5.43) 得证: (5.44)

$$\mathrm{tr}(AB) = \mathrm{tr}(BA). \tag{5.45}$$

为了证明的完备性,我们给出(5.43)的证明: 如果 B 是非奇异矩阵,则

$$xI - AB = B^{-1}(xI - BA)B,$$

那么它们的模(又称矩阵的行列式)相同

$$\begin{aligned} \det(xI - AB) &= \det(B^{-1})\det(xI - BA)\det(B) \\ &= \det(xI - BA), \end{aligned} \tag{5.46}$$

其中

$$\det(B^{-1})\det(B) = \det(B^{-1}B) = \det(I) = 1.$$

如果矩阵 B 是奇异矩阵, 设ϵ 为一个实数, 根据定理如果要让$B - \epsilon I$ 矩阵也是奇异矩阵当且仅当 ϵ 是矩阵 B的特征值.而根据定理,$n \times n$的矩阵 B 最多有n 个特征值, 当ϵ不等于任意一个特征值时会使$B - \epsilon I$ 为非奇异矩阵. 当$B - \epsilon I$为非奇异矩阵时根据 (5.46)有:

$$\det(xI - A(B - \epsilon I)) = \det(xI - (B - \epsilon I)A). \tag{5.47}$$

因为 (5.47)的两边均为关于ϵ的 n 阶多项式, 如果让$\epsilon \to 0$ 时(5.47)会变形为

$$\begin{aligned} & \lim_{\epsilon \to 0} (\det(xI - A(B - \epsilon I))) \\ = \quad & \lim_{\epsilon \to 0} (\det(xI - (B - \epsilon I)A)) \\ \Longleftrightarrow \quad & \det(xI - AB) = \det(xI - BA). \end{aligned} \tag{5.48}$$

根据 (5.46) 和 (5.48), 在任意情况下都有如下的关系: $\det(xI - AB) = \det(xI - BA)$, 从而证明了等式 (5.43). □

题目 8. 能否找到满足以下关系的 $n \times n$ 矩阵 A 和 B?

$$AB - BA = I_n,$$

其中 I_n 是$n \times n$的单位矩阵.

参考答案: 使用反证法证明: 假如存在 $n \times n$ 的两个矩阵 A 和 B 满足 $AB - BA = I$, 则

$$\mathrm{tr}(AB - BA) = \mathrm{tr}(I_n) = n. \tag{5.49}$$

又因为A 和 B 为$n \times n$的矩阵,根据 (5.45)有: $\text{tr}(AB) = \text{tr}(BA)$.

$$\text{tr}(AB - BA) = \text{tr}(AB) - \text{tr}(BA) = 0, \tag{5.50}$$

与原假设矛盾.所以不存在这样的两个矩阵 A 和 B 使得$AB - BA = I$. □

题目 9. 概率矩阵(probability matrix)是一个所有元素非负且每行之和都为1的矩阵. 请证明两个概率矩阵的乘积仍为概率矩阵.

参考答案: 把上面的话用数学公式表达出来:

设 $n \times n$ 矩阵 M 为概率矩阵,那么它的所有元素非负并且

$$M\mathbf{1} = \mathbf{1}, \tag{5.51}$$

其中 $\mathbf{1}$ 为 $n \times 1$的列向量 ,它的元素都是1.

将 $M = \text{row}\,(r_j)_{j=1:n}$ 用行向量表示, 其中r_j 为$1 \times n$的行向量. 如果把第j行的所有元素 r_j 加起来就有[4]

$$\sum_{k=1}^{n} r_j(k) = r_j\mathbf{1}. \tag{5.52}$$

概率矩阵的每一行之和为1的条件可以表达成:

$$\begin{aligned}
& \sum_{k=1}^{n} r_j(k) = 1, \ \forall\, j = 1:n \\
\Longleftrightarrow \quad & r_j\mathbf{1} = 1, \ \forall\, j = 1:n \\
\Longleftrightarrow \quad & (r_j\mathbf{1})_{j=1:n} = \mathbf{1} \\
\Longleftrightarrow \quad & M\mathbf{1} = \mathbf{1},
\end{aligned}$$

其中 $M\mathbf{1} = (r_j\mathbf{1})_{j=1:n}$, $M = \text{row}\,(r_j)_{j=1:n}$ 为M的行向量表达形式.

[4]注意 r_j为$1 \times n$ 行向量而 $\mathbf{1}$ 是 $n \times 1$的列向量, 所以 $r_j\mathbf{1}$的表达方式与 (5.52)是一致的.

换句话说如果证明了 (5.51)对于AB也成立就证明了矩阵 M是概率矩阵.

设 A 和 B 为概率矩阵则所有A 和 B 的元素非负,从而AB 所有元素也非负. 根据 (5.51)有:

$$A\mathbf{1} = \mathbf{1} \quad \text{和} \quad B\mathbf{1} = \mathbf{1},$$

那么

$$(AB)\mathbf{1} \;=\; A(B\mathbf{1}) \;=\; A\mathbf{1} \;=\; \mathbf{1}.$$

再根据 (5.51), 得证 AB 也是概率矩阵. □

题目10. 求 ρ的取值范围使得

$$\begin{pmatrix} 1 & 0.6 & -0.3 \\ 0.6 & 1 & \rho \\ -0.3 & \rho & 1 \end{pmatrix}$$

是相关系数矩阵.

参考答案: 这道题目的答案可以通过西尔维斯特准则 (Sylvester's criterion) 得到.详细的参考答案请回到章节 1. 除此之外我们提供另外两种解法供大家参考:一种是使用楚列斯基矩阵分解的方法,另外一种方法是基于对称半正定矩阵的定义.

一个矩阵是相关系数矩阵的充要条件为:它是对称半正定矩阵并且它的的对角元素全部为 1.因此我们要寻找这样的 ρ 使得

$$\Omega \;=\; \begin{pmatrix} 1 & 0.6 & -0.3 \\ 0.6 & 1 & \rho \\ -0.3 & \rho & 1 \end{pmatrix} \tag{5.53}$$

为对称半正定矩阵.

解法 1: 为了寻找使得Ω成为对称半正定矩阵的ρ, 对Ω使用楚列斯基分解,得到了 2×2 的矩阵:

$$
\begin{aligned}
& \begin{pmatrix} 1 & \rho \\ \rho & 1 \end{pmatrix} - \begin{pmatrix} 0.6 \\ -0.3 \end{pmatrix} (0.6 \;\; -0.3) \\
= & \begin{pmatrix} 1 & \rho \\ \rho & 1 \end{pmatrix} - \begin{pmatrix} 0.36 & -0.18 \\ -0.18 & 0.09 \end{pmatrix} \\
= & \begin{pmatrix} 0.64 & \rho + 0.18 \\ \rho + 0.18 & 0.91 \end{pmatrix}.
\end{aligned}
$$

那么使Ω 为半正定矩阵的充要条件是矩阵

$$M = \begin{pmatrix} 0.64 & \rho + 0.18 \\ \rho + 0.18 & 0.91 \end{pmatrix}$$

为半正定矩阵.又因为 $M(1,1) = 0.64 > 0$, 所以当且仅当$\det(M) \geq 0$时M 为半正定矩阵, 即

$$\det(M) = 0.5824 - (\rho + 0.18)^2 \geq 0, \tag{5.54}$$

简化为

$$|\rho + 0.18| \leq \sqrt{0.5824} = 0.7632.$$

综上所述: 当且仅当

$$-0.7632 \leq \rho + 0.18 \leq 0.7632,$$

时 Ω 为半正定矩阵.又Ω是对角元素均为1的对称矩阵,因此它是相关系数矩阵. 此时ρ的取值范围是:

$$-0.9432 \leq \rho \leq 0.5832. \tag{5.55}$$

答案与使用西尔维斯特准则(Sylvester's criterion)得到的结果相同. 详细过程详见章节(1.9) .

解法 2: 根据定义,矩阵 Ω为对称半正定矩阵的充要条件是$x^t\Omega x \geq 0$ 其中 $x = (x_i)_{i=1:3} \in \mathbb{R}^3$ 注意到

$$\begin{aligned} & x^t\Omega x \\ = & (x_1\ x_2\ x_3)\begin{pmatrix} 1 & 0.6 & -0.3 \\ 0.6 & 1 & \rho \\ -0.3 & \rho & 1 \end{pmatrix}\begin{pmatrix} x_1 \\ x_2 \\ x_3 \end{pmatrix} \\ = & x_1^2 + x_2^2 + x_3^2 + 1.2x_1x_2 - 0.6x_1x_3 + 2\rho x_2x_3. \end{aligned}$$

凑成平方的形式则会得到

$$\begin{aligned} & x^t\Omega x \\ = & x_1^2 + 2x_1(0.6x_2 - 0.3x_3) + x_2^2 + x_3^2 + 2\rho x_2x_3 \\ = & (x_1 + 0.6x_2 - 0.3x_3)^2 \\ & -(0.6x_2 - 0.3x_3)^2 + x_2^2 + x_3^2 + 2\rho x_2x_3 \\ = & (x_1 + 0.6x_2 - 0.3x_3)^2 \\ & +0.64x_2^2 + 2x_2x_3(\rho + 0.18) + 0.91x_3^2. \end{aligned}$$

再次凑平方:

$$\begin{aligned} & 0.64x_2^2 + 2x_2x_3(\rho + 0.18) + 0.91x_3^2 \\ = & \left(0.8x_2 + x_3\frac{\rho + 0.18}{0.8}\right)^2 \\ & -x_3^2\frac{(\rho + 0.18)^2}{0.64} + 0.91x_3^2 \\ = & \left(0.8x_2 + x_3\frac{\rho + 0.18}{0.8}\right)^2 \\ & +\frac{x_3^2}{0.64}\left(0.5824 - (\rho + 0.18)^2\right), \end{aligned}$$

从而

$$
\begin{aligned}
& x^t \Omega x \\
= \; & (x_1 + 0.6x_2 - 0.3x_3)^2 \\
& + \left(0.8x_2 + x_3 \frac{\rho + 0.18}{0.8}\right)^2 \\
& + \frac{x_3^2}{0.64}\left(0.5824 - (\rho + 0.18)^2\right).
\end{aligned}
$$

所以使得$x^t \Omega x \geq 0$ $x = (x_i)_{i=1:3} \in \mathbb{R}^3$ 的充要条件是:

$$
\begin{aligned}
& (x_1 + 0.6x_2 - 0.3x_3)^2 \\
& + \left(0.8x_2 + x_3 \frac{\rho + 0.18}{0.8}\right)^2 \\
& + \frac{x_3^2}{0.64}\left(0.5824 - (\rho + 0.18)^2\right) \\
\geq \; & 0, \; \forall \, x_1, x_2, x_3 \in \mathbb{R}.
\end{aligned}
$$

简化之后得到:

$$
0.5824 - (\rho + 0.18)^2 \; \geq \; 0, \tag{5.56}
$$

这与 (5.54)的结果相同

所以我们得出以下结论,当且仅当

$$
-0.9432 \; \leq \; \rho \; \leq \; 0.5832
$$

时 Ω 是相关系数矩阵 □

5.3 金融产品: 期权,债券,掉期,远期,期货.

题目1. 已知三个欧式认沽期权合约的行权价格分别为 40, 50和70,其余合约内容相同(标的物相同,到期时间相同). 请问如果它们的市场售价分别为\$10, \$20和\$30, 是否有无风险套利的机会. 如果有, 请详述套利策略.

参考答案: 如果认沽期权的"凸性"被破坏了, 则一定存在无风险套利机会. (凸性为看跌期权和看涨期权价格对于标的物价格的二阶导数大于0).

在(K, y)平面上(K为期权的行权价, y为期权的价格), 通过点$(K = 40, P(40) = 10)$和点$(K = 70, P(70) = 30)$的直线可以写作

$$y = \frac{70 - K}{30} \cdot 10 + \frac{K - 40}{30} \cdot 30. \tag{5.57}$$

那么 $K = 50$的点在这条直线(5.57)上所对应的y值为

$$\frac{2}{3} \cdot 10 + \frac{1}{3} \cdot 30 = \frac{50}{3}.$$

根据凸性,期权价格$P(K)$是关于行权价格K的一个凸函数. $K = 50$的期权合约价格应当低于直线上对应的$K = 50$处的投资组合价值.然而题目中给出的 $K = 50$的期权合约价格$P(50) = 20 > \frac{50}{3}$,所以存在无风险套利机会.

为了抓住这个无风险套利机会, 我们采用"高买低卖"的策略: 买2手行权价为40的认沽期权, 买1手行权价为70的认沽期权. 同时卖出3手行权价为50的认沽期权. 构建上述投资组合可以为我们带来\$10收益:

$$3 \cdot \$20 - 2 \cdot \$10 - \$30 = \$10.$$

设$V(T)$为到期(T)时以上投资组合的价格,则

$$\begin{aligned} V(T) &= 2\max(40 - S(T), 0) \\ &\quad + \max(70 - S(T), 0) \\ &\quad - 3\max(50 - S(T), 0). \end{aligned}$$

注意到不管标的物价格$S(T)$取何值, $V(T)$均非负:

如果到期时$70 \leq S(T)$,所有期权合约都变成价外期权,则

$$V(T) = 0.$$

如果到期时 $50 \leq S(T) < 70$, 则

$$V(T) = 70 - S(T) \geq 0.$$

如果到期时$40 \leq S(T) < 50$, 则

$$\begin{aligned} V(T) &= (70 - S(T)) - 3(50 - S(T)) \\ &= 2S(T) - 80 \\ &\geq 0. \end{aligned}$$

如果到期时 $S(T) < 40$, 则

$$V(T) = 2S(T) - 80 + 2(40 - S(T)) = 0.$$

换句话说我们配置出了一个投资组合,无论标的物的价格如何变化,这个投资组合在到期时的回报均非负.其无风险收益为初始的\$10在到期时的价值. □

题目 2. 假设一公司的股票价格为 \$50,并传闻在三个月之内将会被另外一家公司兼并收购,如果成功收购,这家公司股票将会上升至\$52,如果失败则会下跌至 \$47,并且成功失败的概率各为50% .有一个以此公司股票为标的物的三个月平值认沽期权,并假设此公司不分红且无风险利息为0.问平值期权合理的价格为多少?

参考答案: 需要牢记的是,真实世界中的概率不会影响期权的二叉树定价. 因此本题在真实世界中收购成功的概率为无用信息,是烟幕弹.

解 1: 在风险中性测度原则下(译者注:即假设某个理想的世界里不存在无风险套利机会,即所有确定收益资产的价格以无风险利率r增长,并存在一系列基本证券,其中某一类基本证券定义为在事件E发生时投资者收益\$1而其他情况下收益为\$0.没有无风险套利机会.已知某类基本证券起始价

格为$p(0)$, 设E发生概率为N, 则它在到期前的价格为$p(1) = N\times1+(1-N)\times0$,根据市场有效性假设, $p(1) = p(0)\times e^{rt}$.所以在已知t,r 和$p(0)$时可以算出风险中性测度下的概率N,反之如果已知 N也可以计算出此类基本证券的现价$p(0)$)由于题目假设无风险利率为0,则起始时的认沽期权价格与到期前价格相同. 用公式表示为:

$$P(0) \;=\; p_{RN,up}P_{up} \;+\; p_{RN,down}P_{down}. \tag{5.58}$$

(译者注:根据读者的需求,此处给出求风险中性概率方法:股票可以通过购买52个A类基本证券和47个B类基本证券等价构造.其中A类基本证券只有在兼并成功时为投资者产生$1收益,B类基本证券只在兼并失败时为投资者产生 $1收益.已知这一篮子基本证券在$t=0$时价值为$50,并且这两个事件互斥.根据译者之前的注释,可以很容易得出两个事件在风险中性原则下发生的概率).

同时设在风险中性测度下$p_{RN,up}$为股票上涨的概率(兼并成功),$p_{RN,down}$为股票下跌的概率(兼并失败),则有:

$$\begin{cases} p_{RN,up}\cdot 1\cdot 52 + p_{RN,down}\cdot 1\cdot 47 \;=\; 50 \\ p_{RN,up} + p_{RN,down} \;=\; 1 \end{cases}$$

解得:

$$p_{RN,up} = 0.6; \quad p_{RN,down} = 0.4.$$

对于认沽期权,已知当股票下跌到$47时价格为$P_{down} = 3$,股票上涨到 $52时价格为 $P_{up} = 0$.根据 (5.58),当无风险利率为0时平值认沽期权的现价为

$$P(0) \;=\; 0.6\cdot 0 + 0.4\cdot 3 \;=\; 1.2. \tag{5.59}$$

则平值认沽期权的价格应当为 $1.20.(译者注:如果真实世界中兼并成功的概率为50%则股票没有在期望价格交易,存在无风险套利机会)

解 2: 对于只有两互斥结果的世界里,可以用现金和股票来构成一个与期权合约等价的投资组合. 其中股票的权重(即

构成投资组合的股票数目)为Δ_P:

$$\Delta_P = \frac{P_{up} - P_{down}}{S_{up} - S_{down}} = \frac{0-3}{52-47} = -0.6.$$

如果购入平值认沽期权并看空上述投资组合从而构成一个更大的投资组合设为Π,(组合中空仓Δ_P 股股票,购入一手三个月平值认沽期权,并且借入使投资组合Π起始价值为0的现金.) 则投资组合Π在任何情况下到期前的价值都为0.证明如下:

$$\begin{aligned}\Pi(T) &= P(T) - [\Delta_P(S(T) - S(0)) + C] \\ &= \begin{cases} 0 - [\Delta_P \cdot (52-50) + C] = 0, \text{ if } S(T) = 52; \\ 3 - [\Delta_P \cdot (47-50) + C] = 0, \text{ if } S(T) = 47. \end{cases}\end{aligned}$$

解出Δ_P和C 分别为-0.6 和1.2之后代入原式则有

$$\Pi(0) = \Pi(T) = 0 = P(0) - [-0.6 \cdot (50-50) + 1.2] = 0,$$

解出$P(0) = 1.2$;与方法1(5.59)中得到的答案相符. □

题目3. 有一支现价为 \$50的股票,在三个月之后等概率变成 \$60 或者 \$40. 并有一个以这支股票为标的物的三个月平值(ATM)认沽期权,价值 \$4. 问这个认沽期权的价格在下列情况中价格如何变化? 股票上涨到 \$60 的概率为 75% 而跌到 \$40 的概率为 25%.

参考答案: 这里仍然可以使用二叉树模型,如同上一题中所述,实际世界中的概率根本不会影响期权的定价.因此即便股票上涨到 \$60 的概率变为 75%,三个月看涨期权的价格仍然为 \$4. □

题目 4. 什么是风险中性定价?

参考答案: 风险中性定价法是描述将衍生产品到期日的期望(平均)收益折现到当前时刻作为衍生物现价的方法.一般会假设标的物在到期日的价格满足以无风险利率为漂移项的对数正态分布.进一步说明,如果一个标的物到期日的价格

满足对数正态分布,(译者注:设标的物起始价格为S_0,如果满足漂移项为无风险利率r,波动率为σ的对数正态分布.则t时刻标的物的对数价格 $\ln(S_t)$的增长率服从$N(rt,\sigma^2 t)$正态分布) 假设标的物分红的回报率为q,且衍生品在到期日T时的期望(平均)收益为 $V(T)$,根据风险中性定价原理,当前时刻的价格可以写成如下形式:

$$V(0) = e^{-rT} E_{RN}[V(S(T))],$$

其中 r 是无风险利率, $S(T)$为标的物在到期日的价格.根据对数正态分布的性质可以得到:

$$S(T) = S(0)\, e^{\left(r-q-\frac{\sigma^2}{2}\right)T + \sigma\sqrt{T}Z}.$$

风险中性定价原理一般适用于“路径无关” 的衍生品定价,例如欧式期权, 或者简单的定点期权(即如果到期日时标的物价格高于行权价,收益为固定的常数或1股标的物股票).这类衍生品的价格只与到期日时标的物的价格有关. 值得注意的是,风险中性原则不可以用于美式期权,障碍期权或者亚洲式期权的定价(这些衍生品的价格跟标的物的整条价格“路径” 有关).

□

题目 5. 简单地描述如何推导BS模型(Black-Scholes formula).

参考答案: BS模型用于对欧式期权进行定价.它假设标的物到期时的价格分布服从对数正态分布.以下为推导BS模型的方法.

- 风险中性定价: 期权在到期日时的平均价格可以通过以下方式计算.首先假设标的物到期时的价格分布为漂移项是无风险利率的对数正态分布.根据不同标的物价格和价格的概率分布,可以计算出到期时期权的期望(平均)价格.(译者注:由于已知标的物到期时的价格分布,对于标的物到期时每一个可能的价格我们都知道它的概率.同时每一个标的物到期时的价格根据期权收益曲线可以得到对应的期权收益.则期权价格在到期日的期望就很可以很容易得到.)

• BS偏微分方程的解: 期权的价格符合BS偏微分方程(Black–Scholes PDE).并且根据期权合约到期时的收益曲线可以得到偏微分方程的边界情况.由于我们在大学里最熟悉的偏微分方程是热传导偏微分方程(Heat PDE),所以需要对BS偏微分方程进行变量替换,将它转变成热传导偏微分方程的形式从而应用公式求出BS偏微分方程的解,得到期权的定价.

• 二叉树定价模型: 由于当二叉树的时间间隔无限接近0时,这种二叉树所对应的二项分布会无限逼近于正态分布.根据风险中性假设可以测定二叉树模型的参数.从而倒推出期权的现价. (译者注:因为二叉树最底层上每一个"叶子"节点代表标的物到期时可能的价格.根据期权的收益曲线,就知道每个"叶子"节点的期权价格. 然后可以反向倒推出每一层节点的期权价格.其中根节点的期权价格就是BS模型下的期权的现价)

注:在这里 Wilmott [4]有兴趣的读者可以找到近12种BS模型的解法. □

题目6. 一个三个月到期的平值认沽期权,标的物价格为$40且标准差为30%.它的价格应当为多少? 为了方便计算,假设无风险利率为0,并且标的物不分红.

参考答案: 当标准差不是很大(例如 $\sigma^2 T \leq 0.25$)且标的物不分红,利率为0的情况下,对于平值认沽期权价格有如下的近似公式:

$$P_{ATM} \approx 0.4\sigma S_0\sqrt{T}; \tag{5.60}$$

详细推导请参考 Stefanica教授编写的 [3] (5.60).

将数字代入上述公式,得到期权价格为2.40.为了验证结果,使用BS模型计算的结果为2.3914.所以近似公式(5.60) 是非常精确的. □

题目7. 如果一个认购期权的标的物价格一日之内增长了一倍,问此期权的价格如何变化?

*参考答案:*首先假设标的物不付股息且无风险利率为r.根据期权的行权价格分情况讨论:当认购期权为实值期权时(行权

价格远低于股价),使用买卖权平价关系(Put-Call Parity),认购期权的价格可以近似写成(译者注:之所以可以用这种方法是因为此时认沽期权价格十分接近0.) $C \approx S - Ke^{-rT}$, 其中 K 和 T 分别为行权价格与合约距到期日时长(译者注:一般以年为单位), r 为无风险利率. 如果股价S增长一倍,认购期权依然为实值期权,所以上述公式仍适用.增长后的价格近似于 $2S - Ke^{-rT}$.我们发现认购期权的价格增长与股票价格(绝对)增长相同.

当此期权为平值期权时,如果标的物价格增长一倍,期权本身将从平值期权变为实值期权.其收益有一个数量级的增长.(译者注:当股票价格变化后,期权价格根据第一种情况可以近似为$2S - Ke^{-rT}$).

当期权为虚值期权时(价格一般为股价的零头),如果标的物价格增长一倍,期权价格一般会增长10倍甚至更多(视情况而定)

假设一认购期权还有6个月到期,行权价格为 20 ,实际波动率为 25%. 设无风险利率为 5%.使用BS定价模型,可以算出对于上述认购期权在标的物处于不同价格时对应期权合约的价格:

标的物价格	认购期权价格
10	0.000045
20	1.65
40	20.49
80	60.49
400	380.49
800	780.49

如果认购期权为虚值期权并且标的物价格从$10 涨到 $20,则期权价格会从$0.000045 涨到 $1.65, 增长了10000倍. 如果是平值期权,标的物股价从 $20 涨到 $40, 期权价格从 $1.65 涨到$20.49,增长了10倍.

对于实值期权而言,如果标的物价格从 \$40涨到 \$80, 期权合约价格从 \$20.49 涨到 \$60.49, 增长了2.95倍. 如果行权价格远小于未涨前的标的物价格,[5] 股票价格从 \$400 涨到\$800, 合约价格从 \$380.49 涨到\$780.49, 仅仅翻了一番.

在上述条件下,如果股票价格大于 \$40,以下近似公式 $C \approx S - Ke^{-rT}$ 可以比较准确的近似期权价格. 那么对于平值期权而言,价格的增长跟标的物的(绝对)增长相同. □

题目8. Delta的取值范围是什么?

参考答案: 设标的物不付股息.对于一个多仓的欧式认购期权,Delta(Δ) 范围是 0 到 1 (对于空仓的欧式认购期权,Delta(Δ) 范围是 −1 到 0). 而且当标的物价格增长时,认购期权的Delta(Δ)从0(此时标的物的价格也为0)到1(此时标的物的价格很大,行权价远小于标的物价格).

对于一个多仓的欧式认沽期权,它的Delta(Δ) 范围是 −1 到 0;如果是空仓的认沽期权,它的Delta(Δ) 范围则是 0 到 1. 多仓认沽期权的Delta随着标的物价格的上升而减小(此处为绝对值减小).即当标的物价格为0时,Delta为−1,当标的物价格很高时(K<<S行权价远小于标的物价格) 它的Delta为0. 根据买卖权平价关系(Put-call parity): $C - P = S - Ke^{-rT}$, 式子两边对标的物价格S求一阶导得到: $\Delta(P) = \Delta(C) - 1$,这个等式映证了上述结论. □

题目 9. 平值认购期权和平值认沽期权的Delta分别是多少?

参考答案: 平值认购期权的Delta近似等于 0.5; 平值认沽期权的Delta近似等于−0.5.

设标的物不付股息,并且无风险利率为0, $\Delta(C_{BS}) = N(d_1)$, 其中 $N(x)$ 为标准正态分布的累积概率方程

$$d_1 = \frac{\ln\left(\frac{S}{K}\right) + \left(r - q + \frac{\sigma^2}{2}\right)T}{\sigma\sqrt{T}} = \frac{\sigma\sqrt{T}}{2} \tag{5.61}$$

[5]实际中这种合约的流动性很差,因为行权价远小于股票价格,其杠杆效应较低.

已知 $K = S$ 并且假设 $r = q = 0$. 对积概率函数$N(x)$ 在0点处进行泰勒展开:

$$N(x) \approx 0.5 + \frac{x}{\sqrt{2\pi}} \approx 0.5 + 0.4x, \tag{5.62}$$

已知 $\frac{1}{\sqrt{2\pi}} = 0.3989 \approx 0.4$ 则

$$\Delta(C_{BS}) = N(d_1) = N\left(\frac{\sigma\sqrt{T}}{2}\right) \approx 0.5 + 0.2\sigma\sqrt{T}.$$

又因为对于大多数期权而言 $0.2\sigma\sqrt{T}$是一个非常小的数字(例如: $0.2\sigma\sqrt{T} \leq 0.1$ 如果隐含波动率小于 50% 并且到期日小于1年时).所以可以近似的认为 $\Delta(C_{BS}) \approx 0.5$.

对于认沽期权 $\Delta(P_{BS}) = -N(-d_1)$,根据(5.61)有: $d_1 = \frac{\sigma\sqrt{T}}{2}$; 又根据 (5.62)可知:

$$\Delta(P_{BS}) = -N\left(-\frac{\sigma\sqrt{T}}{2}\right) \approx -0.5 + 0.2\sigma\sqrt{T},$$

同理有如下的结果: $\Delta(P_{BS}) \approx -0.5$. □

题目10. 什么是买卖权平价关系(Put-Call Parity)? 如何证明其成立?

参考答案: 买卖权平价关系是欧式认购与认沽期权(具有相同的行权价和到期日)之间不存在无风险套利机会的必要条件, 这个关系不依赖于模型假设.

具体的讲,如果买入一手认购期权并同时卖出相同行权价和到期日的一手认沽期权,就等价于买入一单位相同标的物,相同到期日并且交割价格为期权行权价的远期合约(假设标的物不付股息)

使用数学公式表达.设 $C(t)$ 和 $P(t)$ 分别为t 时欧式认购期权和欧式认沽期权的价格(合约到期日为T,行权价为K,标的物不付股息并且t时刻标的物价格为$S(t)$).买卖权平价关系可以写成如下形式:

$$C(t) - P(t) = S(t) - Ke^{-r(T-t)}, \tag{5.63}$$

其中 r 为无风险利率. 如果标的物支付连续的股息q,买卖权平价关系有如下形式:

$$C(t) - P(t) = S(t)e^{-q(T-t)} - Ke^{-r(T-t)}. \quad (5.64)$$

为了简便,一般假设标的物不支付股息. (5.63)

如果考虑如下的投资组合:

- 购入一单位认沽期权;
- 购入一单位标的物;
- 卖出(看空) 一单位认购期权.

则t时投资组合的价格为:

$$V_{portfolio}(t) = P(t) + S(t) - C(t). \quad (5.65)$$

认沽期权到期日的价格为:

$$\begin{aligned} C(T) &= \max(S(T) - K, 0); \\ P(T) &= \max(K - S(T), 0). \end{aligned}$$

则无论到期日时 $S(T) < K$ 亦或是 $S(T) \geq K$, 投资组合的价值均为 K.证明如下:

如果 $S(T) < K$,则 $P(T) = K–S(T)$, $C(T) = 0$.此时投资组合价值为:

$$\begin{aligned} V_{portfolio}(T) &= P(T) + S(T) - C(T) \\ &= (K - S(T)) + S(T) - 0 \\ &= K. \end{aligned} \quad (5.66)$$

如果$S(T) \geq K$, 则 $P(T) = 0$, $C(T) = S(T)–K$. 此时投资组合价值为:

$$\begin{aligned} V_{portfolio}(T) &= P(T) + S(T) - C(T) \\ &= 0 + S(T) - (S(T) - K) \\ &= K. \end{aligned} \quad (5.67)$$

通过 (5.66) 和 (5.67), 可以得到以下结论,即投资组合的价格与标的物价格$S(T)$无关.

$$V_{portfolio}(T) = P(T) + S(T) - C(T) = K, \quad (5.68)$$

根据风险中性原理,投资组合在 t 时刻的价格一定等于 K 折现到 t时刻的折现价. 即

$$V_{portfolio}(t) = Ke^{-r(T-t)}. \tag{5.69}$$

通过 (5.65) 和 (5.69), 我们得到

$$P(t) + S(t) - C(t) = Ke^{-r(T-t)}$$

$$C(t) - P(t) = S(t) - Ke^{-r(T-t)}.$$

得证买卖权平价关系的数学公式 (5.63).

如果标的物支付股息,买卖权平价关系 (5.64) 也可以用相同的方法得到. 假设标的物支付连续的股息记做 q ,可以构建如下的投资组合:买入一单位认沽期权,卖出一单位认购期权,并且买入$e^{-q(T-t)}$单位的标的物.假设获得的股息立即以现价买入等值的股票.则到期日时累计持有一单位标的物(股票).根据上述公式,我们知道在 T时,投资组合的价格一定等于 K, 与到期日时的标的物价格 $S(T)$无关.则可以得到在假设标的物支付股息情况下的买卖权平价关系数学公式 (5.64) . □

题目11. 证明欧式认购期权在平值时时间价值最大.

参考答案: 期权的时间价值定义为期权的价格 $C(S)$ 减去其内在价值(对于认购期权为$S-K$, 对于认沽期权为$K-S$) $\max(S-K,0)$.

可写成:

$$C(S) - \max(S-K,0).$$

可以通过证明下面的方程在$S=K$处取得最大值来证明期权的时间价值在 $S=K$时最大.设时间价值方程为[6] :

$$f(S) = C(S) - \max(S-K,0)$$

$$f(S) = \begin{cases} C(S), & \text{if } S \le K; \\ C(S) - S + K, & \text{if } S > K. \end{cases}$$

[6]方程为连续方程, 但$f(S)$ 在$S=K$处不可导.

当 $S \leq K$时

$$f'(S) = \Delta(C) > 0,$$

时间价值方程随着K递增.
当$S > K$时

$$f'(S) = \Delta(C) - 1 < 0,$$

因为$\Delta(C)$小于1[7] . 时间价值方程也随着K递减.
综上所述,时间价格方程 $f(S)$ 在$S = K$时取得最大值.
同理,认沽期权时间价值可以写成:

$$P(S) - \max(K - S, 0),$$

在 $S = K$时取得最大值. 如果设 $g(S) = P(S) - \max(K - S, 0)$, 则

$$g(S) = \begin{cases} P(S) - K + S, & \text{if } S \leq K; \\ P(S), & \text{if } S > K. \end{cases}$$

因为认沽期权的Delta永远大于 -1 且小于0 (当$S \ll K$时取值为-1, 当K≪S时取值为0).所以当$S \leq K$时,

$$g'(S) = \Delta(P) + 1 > 0,$$

时间价值方程$g(S)$ 关于行权价格 K单调递增. 当 $S > K$时,

$$g'(S) = \Delta(P) < 0,$$

时间价值方程 $g(S)$ 关于行权价格 K单调递减. 所以方程在$S = K$处取得最大值.

---

[7]对于多仓的认购期权,随着标的物价格从0到∞ ,delta也从0变到1. 使用BS模型证明,

$$\Delta(C_{BS}) = e^{-qT} N(d_1) < N(d_1) < 1.$$

综上所述,无论认沽或者认购期权,它们的时间方程 $g(S)$ 均在 $S = K$取得全局最大值. □

题目12. 什么是隐含波动率(implied volatility)?
波动率微笑(volatility smile)?
波动率偏离(volatility skew)?

参考答案: 根据定义,隐含波动率 σ是标的物在对数正态分布假设下使根据BS模型计算出的期权价格等于市场价格的唯一波动率参数.(译者注:由于BS模型中大部分变量均可从市场上观测得到,唯一无法直接得到的变量是标的物价格的波动率.通过隐含波动率,人们可以比较不同行权价和不同到期日的期权合约是否被高估或者低估). 值得注意的是,每一个期权合约的市场价对应唯一的隐含波动率. [8]

同样标的物的不同期权由于到期日和行权价格的不同,它们交易的价格也不尽相同. 对于每一类期权,在对数正态分布假设下,根据它市场上交易的价格,通过BS模型可以求出它的隐含波动率. (译者注:相同标的物的期权合约类型可以由1.行权价格2.到期日,唯一确定.且每一类期权合约都对应唯一一个隐含波动率参数.如果把行权价格和到期日分别作为坐标系的X,Y,轴.如果将X,Y 坐标确定的坐标点代表某一类期权.并将这个坐标点标注在以它对应的的隐含波动率大小作为z轴的三维空间里.当市场上流动性非常好时, 这些点非常密集的组成了一个曲面,这就是隐含波动率曲面(volatility surface)).如果我们按照市场上约定俗成的几个到期日进行分类,在每一个相同的到期日下不同行权价格的隐含波动率就会形成一条曲线,这条曲线在实际中一般不是水平直线.但都会有某些共性,主要有以下两点:

第一点,对于相同到期日的期权合约.行权价远小于标的物价格处($K \ll S$)的隐含波动率或者行权价远大于标的物价格处($K \gg S$)的隐含波动率一般均高于平值处($S = K$)的隐含波动率. 如果把隐含波动率按照行权价格从小到大从左到右在二维平面中排列出来,会观察到一条从高到低再变高的曲线,非常像一个微笑. (译者注:谁说数学家没有审

[8]隐含波动率的唯一性是因为期权价格对隐含波动率是严格单调递增函数.

美) 这种现象被称为隐含波动率微笑(volatility smile).(译者注:当隐含波动率对于行权价的二阶导大于0,会有隐含波动率微笑现象,这种现象有的标的物存在,有的却不存在)

第二点,对于相同到期日的期权合约.当隐含波动率曲线在$K = S$附近的斜率不为0时,称为波动率偏离(volatility skew).当斜率为负数时,称为正偏离(forward skew)(一般股指期权,个股期权会出现这种现象). 反之如果斜率为正数时,称为逆偏离(reverse skew).(在商品期货作为标的物的期权合约市场中非常常见)

□

题目13. 什么是期权的$Gamma$?为什么$Gamma$越小越容易控制风险?为什么欧式期权的$Gamma$大于0?

参考答案: 期权的$Gamma(\Gamma)$ 是描述Delta对于标的物价格变化的灵敏程度.

$$\Gamma(V) = \frac{\partial \Delta}{\partial S} = \frac{\partial^2 V}{\partial S^2},$$

其中 V 为期权的价格.

如果期权价格对于标的物价格变化不敏感(Delta较小),它的风险较小.为了规避风险,投资者往往通过交易标的物股票使期权仓位接近Delta中性(Delta neutral).对于某一时刻Delta中性的期权-标的物投资组合,当标的物价格变化时,期权Gamma越小,其Delta变化越小,投资组合整体的Delta越接近于0.投资组合风险变化越小,所需对冲的频率较低.所以对于成熟的投资者而言,一般追求其投资组合的Delta和Gamma均接近于0(也称为Delta neutral,Gamma neutral).值得注意的是,当标的物价格出现断崖式变化时(jump in the price) 这种投资组合仍需要再次对冲达到风险中性.

Gamma的数学定义是Delta对于标的物价格的导数,因为Delta随着标的物价格增加而变大,所以Gamma一定为正数.并且当行权价格接近标的物现价时Gamma较大,远离标的物现价(大于或小于)时Gamma较小. □

题目14. 欧式认购期权的价格与欧式认沽期权的价格在什

么情况下相同? (上述认购与认沽期权的行权价格,标的物和到期日均相同) 如何直观地理解本题的结论?

参考答案: 根据买卖权平价关系(Put-Call parity)

$$C - P = Se^{-qT} - Ke^{-rT}, \tag{5.70}$$

其中 C 和 P 分别代表行权价为K,到期时间为T 的认购和认沽期权合约的价格. 并设标的物价格为S 且标的物连续支付股息记做q. 令$C = P$, 则(5.70) 可以简化为 $K = Se^{(r-q)T}$. 由于标的物的远期价格为 $F = Se^{(r-q)T}$, 所以当行权价格等于标的物远期价格时,认购期权和认沽期权价格相同, 这种期权称为远期平值期权(at–the–money–forward options).

因为买卖权平价关系(Put-Call parity)不是建立在任何模型假设之上,所以上述结论在任何模型假设下均成立.

同学们往往会根据直觉认为,远期平值认购期权(at–the–money–forward call)与远期平值认沽期权 (at–the–money–forward put) 的市场价格不会相同, 因为根据认购期权和认沽期权的定义:

$$C(T) = \max(S(T) - K, 0),$$

认购期权的最大收益是无限大.

$$P(T) = \max(K - S(T), 0),$$

而因为股价的不会小于0,所以认沽期权的最大收益是有限的.然而实际上价格等于0和价格等于无穷大的概率均为无限小,并且价格上涨下跌相同幅度的概率不相同.从而导致了在最大收益不同的情况下认购期权与认沽期权价格可以相同.举一个例子,当使用风险中性原理定价时期权价格由合约到期日的平均收益决定.对于所有常见的模拟/描述标的物价格变化的模型(例如几何布朗运动模型),标的物极端价格的概率密度是成指数递减,使标的物取到两个极值的概率接近0.如果画出回报率的概率密度分布图,会发现负半轴尾部概率比正半轴远端的概率要“肥”大(称为fat tail),即变化相同幅度时下跌比上涨的概率大.这是因为当大跌时会引起市场恐慌从而加速股价进一步下跌.所以由于概率的偏移导致了在绝对收益不对称的情况下,远期平值认购期权和远期平值认沽期权的价格相同.

□

题目15. 如果一个标的物的六个月波动率为30%,那么它两年的波动率为多少?

参考答案: 波动率(资产回报率的标准差)大小是以时间的平方根为量级进行缩放的.例如记 σ 为资产价格的年化波动率,并记$\sigma(t)$ 为资产在时间 t内的波动率,则: $\sigma(t) = \sigma\sqrt{t}$. 即

$$\frac{\sigma(t_2)}{\sigma(t_1)} = \sqrt{\frac{t_2}{t_1}}.$$

把原题的条件代入上式 $t_2 = 2$, $t_1 = 0.5$ 并且 $\sigma(t_1) = \sigma(0.5) = 0.3$, 得到 $\sigma(t_2) = \sigma(2) = 0.6$, 则两年的波动率为60%.(译者注:一般不做特殊说明时,期权的隐含波动率或实现波动率均是年化波动率.这是因为BS模型中的时间是以年为单位) □

题目16. 利率掉期(interest rate swap)是如何定价的?

参考答案: (译者注:掉期(swap)英文的字面意义是“交换”.掉期合约本质是让参与合约的双方进行现金流交换来达到交换风险的目的.设合约的面值为100万元,参与合约的双方A与B.合约规定付息日时,A定期付给B以固定利率计算的现金流,B则定期回付以现时浮动利率计算的现金流.其中浮动利率往往参考某个公认的标准利率例如伦敦银行间隔夜拆借利率LIBOR) 掉期合约有两组现金流,合约双方仅仅在付息日时支付对方相应利息,在到期日时不进行面额交换.为了方便计算,可以假设在到期日时,双方进行了数量为面值的等额现金交换.这虚构的现金流分别加上合约规定的两组利息现金流就等价于两种债券.一种是固定利率债券,一种是浮动利率收益债券.那么掉期的合约价格为两种债券折现价格之差: $V_{swap} = V_{fix} - V_{float}$, 其中 V_{fix}是固定利率债券折现价,它的固定利率与掉期合约中的固定利率相同. V_{float}为浮动利率债券折现价格,其浮动利率与掉期合约中规定的浮动利率相同.

首先计算 V_{fix},它是标准的固定收益债券,把所有现金流折现到当前时间就得到了它的价格(一般使用伦敦银行间同业拆借利率 LIBOR 为折现利率).

接着计算V_{float}浮动利率债券价格.当浮动利率等于折现率时,浮动利率债券在发行时的价格等于面值. 这是因为折现率为市场平均期望利率,即投资者持有现金在一段时间内获得的市场平均利率.那么当浮动利率与折现率相同时,浮动利率债券到期日时累计现金流与持有面值等额现金收取市场平均利率所累计的利息现金流相同.即购买这样的浮动利率债券与一开始持有面值等额的现金等效.所以每次付息日结束时,浮动利率债券价格等于累计利息加上面值.

举一个例子:设有一面值为$1千万元,19个月到期的掉期.它的固定利率为3%,折现率与浮动利率同为伦敦同业拆借利率(LIBOR)并使用利滚利(compounded)计算方法,每半年付一次利息.已知下一次浮动利率的现金流将是$125,000(一般是在支付利息日的5个月前确定下来的). 合同规定付息时间为合约签署后的第1, 第7, 第13 和19 个月. 对于支付固定利率的一方,掉期的价格为 $V_{swap} = V_{fix} - V_{float}$. 固定利率现金流的折现价格为:

$$\begin{aligned} V_{fix} \quad = \quad & 150,000 \cdot \text{disc}\left(\frac{1}{12}\right) \\ & + 150,000 \cdot \text{disc}\left(\frac{7}{12}\right) \\ & + 150,000 \cdot \text{disc}\left(\frac{13}{12}\right) \\ & + 10,150,000 \cdot \text{disc}\left(\frac{19}{12}\right), \end{aligned} \tag{5.71}$$

$\text{disc}(t)$ 记做时间 t时的折现系数. 浮动现金流部分.由于每半年付一次利息则它的折现率为:

$$\text{disc}(t) = \left(1 + \frac{\text{LIBOR}(t)}{2}\right)^{-2t}.$$

根据上述计算浮动利率价格的方法,浮动利率债券在付息日结束时的价格为累计持有的利息加上面额.投资者下一付息日结束时累计持有现金为 \$125,000 加上面值 \$1千万共计 \$10,125,000.

则有

$$V_{float} = 10,125,000 \cdot \text{disc}\left(\frac{1}{12}\right). \tag{5.72}$$

那么 $V_{swap} = V_{fix} - V_{float}$, 其中$V_{fix}$ 和 V_{float}分别由 (5.71) 和 (5.72)计算而得. □

题目17. 如果债券收益率上涨10个基准点,10年无息债券(zero coupon bond)的价格变化多少?

参考答案: 记债券价格变化为ΔB,折现率的变化为 Δy,则有

$$\Delta B \approx \frac{\partial B}{\partial y}\Delta y = -DB\Delta y, \tag{5.73}$$

其中 $D = -\frac{1}{B}\frac{\partial B}{\partial y}$ 一般称为债券的久期(duration) .则:

$$\frac{\Delta B}{B} \approx -D\Delta y. \tag{5.74}$$

对于无息债券而言,它的久期等于它的到期时间, $D = 10$. 又已知$\Delta y = 0.001$,根据 (5.74),债券价格变化为:

$$\frac{\Delta B}{B} \approx -10 \times 0.001 = -0.01.$$

则如果平均回报率(折现率)上涨十个基本点,十年期无息债券的价格将会下降 1% . □

题目18. 一个久期(duration)为3.5年的5年期债券的价格为102.如果收益率(yield) 下降50个基点,这个债券的价格变化多少?

参考答案: 因为收益率下降,所以债券价格会上升.并且根据 (5.73) 有:

$$\Delta B \approx -DB\Delta y. \tag{5.75}$$

其中债券价格变化为ΔB,市场期望收益率的变化为Δy,久期为D并且债券价格为B. 已知 $B = 102$, $D = 3.5$ 并

且 $\Delta y = -0.005$ 通过 (5.75) 可知:债券价格变化为 $\Delta B \approx 1.785$. 因此折现率下调之后的债券价格为 $B + \Delta B \approx 103.785$. □

题目19. 什么是远期合约(forward contract)? 什么是远期价格(forward price)?

参考答案: 远期合约有两个签约方,多头寸方(the long position)同意与空头寸方(the short position)以某个协议规定的价格在未来的某个时间购买合约规定数目的标的物.而这个合约规定的价格则称为远期价格(forward price)[9] .合约中规定的远期价格会使远期合约的现价等于0.

如果标的物的现价为S_0并且连续地支付股息q. T时刻到期的合约的远期价格应当为

$$F = S_0 e^{(r-q)T},$$

其中 r为无风险利率. □

题目20. 国债期货合约(treasury futures contract)的远期价格应当如何计算?大宗商品期货合约的远期价格又该怎样计算呢?

参考答案: 看空(short)一单位的远期合约(卖出远期合约)可以通过购入相同数量的标的物进行完美的对冲.(译者注:卖出远期合约可以看成是与对手方约定在到期日以合约规定的价格卖给对手方合约规定数量的标的物,如果沽方从签约时就已经购入了标的物,无论标的物价格如何变化,其收益与损失不会变化.这是因为已经持有的标的物,其购买价格确定,售出价格通过卖出远期合约也已经固定,所以其收益损失固定).远期价格则由两个因素决定,一个为标的物的现价,另一个是持有标的物所产生的费用(例如储存石油的费用).

购买国债相当于购买了未来到期日时与面值等额的现金流,和期间所产生的利息. 设S_0 为国债的现货价格,并且设C 为所有利息折现的总和, T为合约的期限,则远期价格F 可以通过下列方法计算$F = (S_0 - C)e^{rT}$.

---

[9]注意远期价格(forward price)不是远期合约的价格(price of the forward).

对于商品期货而言,购买标的物并一直持有到合约到期会产生存储的费用,例如为了对冲黄金远期期货,需要在签约日开始购买相同数目的黄金,一直持有到到期日交付给对手方,但这就会产生存储黄金的费用. 设S_0 为这批黄金的现价,C 为存储黄金的存储费折现价, 则远期价格F 为: $F = (S_0 + C)e^{rT}$, 其中 r 记做无风险利率. □

题目21. 什么是欧洲美元期货(Eurodollar futures)?

参考答案: 欧洲美元期货是离岸美金的期货,其平均回报率为离岸美金利率(往往十分接近伦敦银行间同业拆借利率LIBOR). 其到期日一般为每年的三月,六月,九月和十二月的第三个星期三,但是也有10年到期的合约. 一般其回报率(yield)使用LIBOR利率近似.例如有一个时长为三个月的国库券期货,其平均回报率可以认为为90天LIBOR利率. □

题目22. 远期合约(forward contract)和期货合约(futures contract)的区别有哪些?

参考答案: 最主要的区别为三点,即合约如何构成,合约如何交割与合约如何交易:

- 期货合约是在交易所公开交易的,所以期货合约各项条款都是遵循统一的标准.而远期合约是在场外交易(OTC),它的条款可以根据客户的要求量身定制.

- 期货合约由于在交易所交易,所以它的盈亏每日都会进行结算(Mark to Market). 然而远期合约则仅仅在到期日时结算盈亏.

- 期货合约由于利润每日结算,它的对手方风险(Counterparty Risk)很小, 与之相反,远期合约的对手方风险较大.

- 期货合约的交割日都是遵循一定的规则,例如上一题中的国库券期货仅仅在某几个月才交割. 然而远期期货可以在任意时间交割.

- 期货合约往往是一手交钱一手交货,即交割时往往会产生实物交割,但远期期货合约则一般以交割日时的现价进行现

金交割,即双方根据远期合约交割日时的现价交换利益和损失,一般不产生实物交换. □

题目23. 如果一个投资组合5天的98%VAR为1千万,那么10天的99%VAR为多少?

参考答案: 已知投资组合的 VaR (Value at Risk) 与时间的平方根成正比.假设某投资组合的短期收益呈正态分布,N 天里$C\%$ 的 VaR 记做 VaR(N, C).其中 N 为时间跨度,C为置信水平.则:

$$\text{VaR}(N, C) \approx \sigma_V z_C \sqrt{\frac{N}{252}}\, V(0), \tag{5.76}$$

其中, σ_V 为投资组合年化收益率标准差, z_C 为z阈值,通过 $P(Z \le z_C) = C$ 确定, $V(0)$ 为投资组合现价.

根据 (5.76)有:

$$\text{VaR}(10\text{ 天}, 99\%) \approx \frac{z_{99}\sqrt{10}}{z_{98}\sqrt{5}}\, \text{VaR}(5\text{ 天}, 98\%).$$

又已知 VaR$(5\text{ 天}, 98\%) = \$10,000,000$, 并且 $z_{99} \approx 2.326348$ 和 $z_{98} \approx 2.053749$, 可以得到:

$$\text{VaR}(10\text{ 天}, 99\%) \approx \$16,019,255. \quad \square$$

5.4 C++. 数据结构.

题目1. 在C++语言里,如何声明数组?如何初始化数组?

参考答案: 数组可以被创建在栈(stack)上,也可以被创建在堆(heap)上.

```
//栈上的数组,未被初始化
T identifier[size];

//栈上的数组,被初始化
T identifier[] = initializer_list;

//堆上的数组,未被初始化
T* identifier = new T[size];
```

例:

```
int foo[3];
int bar[] = {1,2,3};
int* baz = new int[3];
```

题目2. 如何得到变量的内存地址?

参考答案: 在变量名前面加上&符号,例如:

```
T var;
T* ptr = &var
```

例:

```
int foo = 1;
int* foo_ptr = &foo;
```

题目3. 如何声明指针数组?

参考答案: 跟声明数组的方式一样,只不过数组的元素为指针类型:

```
T* identifier[size];
T* identifier[] = initializer_list;
T** identifier = new T*[size];
```

例:

```
int a = 1; int b = 2; int c = 3;
int* foo[3];
int* bar[] = {&a, &b, &c};
int** baz = new int*[3];
```

题目4. 如何声明常量指针(Const Pointer),指向常量的指针(Pointer to a Const),和一个指向常量的常量指针(Const Pointer to a Const)?

参考答案:

```
//指向常量的指针
const T* identifier;
T const* identifier;

//常量指针
T *const identifier = rvalue;

//指向常量的常量指针
const T *const identifier = rvalue;
T const *const identifier = rvalue;
```

例:

```
//常量
const int a = 2; const int b = 2;
int c  = 1;

//指向常量的指针:形式1
int const* foo_two;
foo_two = &a; foo_two = &b;
```

```
//指向常量的指针:形式2
const int* foo;
foo = &a; foo = &b;

//常量指针指向c
//必须在常量指针声明的同时初始化
int *const bar = &c;

//指向常量的常量指针
//必须在常量指针声明的同时初始化
const int *const baz = &a;
```

题目5. 如何动态声明数组?

参考答案:

```
T* identifier = new T[size];
T* identifier = nullptr;
T* identifier;

delete[] identifier;
```

例:

```
int *foo = new int[4];
int *bar = nullptr;
bar = new int[4];
```

题目6. 函数签名(function signature)的一般形式是什么?

参考答案:

```
return_type function_name(parameter_list);
```

例:

```
int my_sum(int a, int b);
```

题目7. 如何实现函数参数的引用传递(Pass-by-Reference)?

参考答案:

```
return_type function_name(T & identifier);
```

此时*identifier*就是原变量的别名,与使用原变量等效,可以修改原变量的值.

题目8. 函数参数如何通过常引用传递?

参考答案:

```
return_type function_name(const T & identifier);
```

通过常引用传递的参数在函数内部无法被改变.

题目9. 引用和指针的区别有哪些?

参考答案:

- 指针可以重新指向其他变量,但引用只能被初始化成某个变量的引用,不能再成为其他变量的引用.
- 指针可以指向NULL(在C++11中是`nullptr`),但没有NULL的引用.
- 指针可以被取地址,但引用不可以.
- 指针有指针运算(pointer arthmetic),但引用没有"引用运算".

题目10. 如何设置函数参数的默认值?

参考答案:

```
return_type function_name(T identifier = rvalue)
```

有默认值的变量在函数的变量列表(又称参数列表)中一般排在没有默认值的变量之后,否则无法区分是覆盖了默认值还是对后面无默认值变量的初始化.

题目11. 如何声明函数模板和模板函数(template function)?

参考答案:

```
template<class T>
return_type function_name(parameter_list);

template<typename T>
return_type function_name(parameter_list);
```

使用 class 或者 typename没有区别.

例:

```
template<typename T>
T temp_sum(T a, T b) {return a+b;}

struct Processor{
    int a;
    int apply(int b) {return a+b;}
};

template<class T>
int temp_sum_2(int a, int b) {
    T processor;
    processor.a = a;
    return processor.apply(b);
}

int main(){
    //implicit, foo equals 3
    int foo = temp_sum(1,2);

    //explicit, bar equals 3
    int bar = temp_sum_2<Processor>(1,2);
}
```

题目12. 如何声明函数指针?

参考答案:

```
return_type (*identifier) (list_parameter_types)
```

例:

```
int my_sum(int a, int b) {return a+b;}
int main(){
    int(*p_func)(int,int);
    p_func = & my_sum;

    // foo equals 3
    int foo = p_func(1,2);
}
```

题目13. 如何禁止C++编译器对类对象进行隐式类型转换?

参考答案: 在声明构造函数时使用关键词 `explicit`:

```
explicit Classname(parameter_list)
```

题目14. 描述C++中`static` 的用处.

参考答案: 在函数中使用关键词`static`声明某个变量意味着一旦这个变量被初始化,它的存储空间直到程序结束才会被回收.

在类的定义中用`static`声明成员变量或成员函数,则这个成员变量或成员函数被这个类的所有对象共享.

将某个文件中的一个全局变量声明为`static`意味着这个变量是这个文件的私有变量.

题目15. 能否把静态函数声明成常量函数(const)?

参考答案: 当某个非静态成员函数被声明成`const`,意味着这个成员函数无法改变非`mutable`的成员变量(其原理相当于把`this`作为一个指向常量的指针看待). 但静态成员函数没有`this`的概念,故`const`对其无效.

题目16. C++中，构造函数支持通过初始化列表对成员变量初始化. 在什么情况下必须使用初始化列表?

参考答案: 当我们需要初始化常量成员变量,引用和没有默认构造函数的成员对象时必须使用初始化列表.除此之外,我们都可以使用函数体对成员变量进行初始化. 但通常情况下,使用初始化列表更加高效,这是因为使用初始化列表时,成员对象的构造函数是直接被调用的. 而如果我们在构造函数的函数体中显式初始化成员对象,在此之前成员对象已被默认构造函数初始化,进入函数体之后成员对象再被赋值.

题目17. 什么是拷贝构造函数?如果类中有指针类型的成员变量,我们用默认的拷贝构造函数会出现哪些问题?

参考答案: 拷贝构造函数可以通过已有的实例来构造新的实例.所有类都有默认的复制构造函数,但默认的构造函数仅仅是浅拷贝(shallow copy)已有实例中的每个成员.如果这个类中有指针成员变量,默认构造函数仅仅将指针拷贝到新的类对象中,这样就会有两个对象的成员指针指向同一片内存区域.如果一个对象被析构, 它的成员指针所指的内存区域就会被收回. 另一个对象中的成员指针则会指向非法区域,从而有可能产生访问非法区域的错误.

值得读者注意的是:当函数的输入变量或返回变量是"值传递"(Pass-By-Value)机制时,如果传递的变量为类对象,编译器会自动调用类的拷贝构造函数.由于拷贝构造函数本身是一个函数,如果拷贝构造函数也使用"值传递",编译器会陷入无限递归调用拷贝构造函数的死循环.所以所有的拷贝构造函数都应当使用"引用传递"(Pass-By-Reference)的机制.为了防止"引用传递"时原实例被更改,往往资深的程序员会加上const关键字.

```
ClassName( const ClassName& other );
ClassName( ClassName& other );
ClassName( volatile const ClassName& other);
ClassName( volatile ClassName& other );
```

题目18. 请给出以下代码的输出:

```
#include <iostream>
using namespace std;

class A
{
public:
    int * ptr;
     A()
    {
        delete(ptr);
    }
};

void foo(A object_input)
{
    ;
}

int main()
{
    A aa;
    aa.ptr = new int(2);
    foo(aa);
    cout<<(*aa.ptr)<<endl;
    return 0;
}
```

参考答案:

这段代码的输出取决于使用的编译器.对于有些编译器,可能会出现编译错误.如果希望得到正确的结果,我们应当显式定义拷贝构造函数.

由于foo函数使用"值传递"机制,当我们调用它时,编译器会自动调用A类型的默认拷贝构造函数浅拷贝aa生成临时对象.当我们离开foo函数后,临时对象被编译器收回,但临时对象与aa中的指针指向同一片数据区域,临时对象被析构造成这片数据区域也被收回,此时aa中的指针所指向的区域为非法区域,所以通过aa指针访问的数据是无效数据甚至会导致某些编译器报错.

题目19.如何重载运算符(overload an operator)?

参考答案:

```
type operator symbol (parameter_list);
```

如果在类定义的外部定义运算符,则此运算符是全局运算符.

例子:

```
struct FooClass{int a;};
int operator + (FooClass lhs, FooClass rhs) {
    return lhs.a + rhs.a;
}
```

题目20. 智能指针(smart pointers)是什么?

参考答案: 智能指针是一个类.其在实现所有指针功能(解析地址dereferencing,间接访问indirection,指针运算arithmetic)的同时提供了智能管理资源的功能.

在C++11中有三种智能指针的实现: `shared_ptr`, `unique_ptr`, 和 `weak_ptr`.

例:

```
//shared_pointer 中有一个计数器来计算有多少个指针
//指向数据.当计数器为0时表明没有指针指向数据,这时
//就可以将内存收回
std::shared_ptr<int> foo(new int(3));
std::shared_ptr<int> bar = foo;

//内存仍没被收回
//bar仍然指向数据
foo.reset();

//当bar被释放,没有指针指向数据内存
```

```
//这时内存被收回
bar.reset();
```

题目21. 什么是封装(encapsulation)?

参考答案: 封装是仅仅将接口提供给用户而将运算和实现包装到后台的开发理念.即用户无法看到所有的技术实现细节.封装通常通过访问权限来实现(public, private, protected).

题目22. 什么是多态(polymorphism)?

参考答案: 多态是一种能够使用同一个接口来操作不同类型的对象的开发理念.(译者注:例如有两种类型"实数"、"虚数",我们可以设计一种加法运算,从而使用同样的函数名对两个不同的类型进行加法操作.)

题目23. 什么是继承(inheritance)?

参考答案: 继承是在继承基类所有的成员函数和成员变量的基础上进行功能的拓展. 这种方式被誉为"white-box"(与之相反的是"black-box")重用.往往某些库文件会提供一些基类供使用者进行拓展.

题目24. 什么是虚函数?什么是纯虚函数?通常在什么情况下使用?

参考答案: 虚函数是同一个函数名有不同的实现(implementation)并且在运行时通过虚函数表(virtual table)选择适当的函数实现.比如在一个基类`Foo`中有一个虚成员函数`f`, 它被继承到了继承类`FooChild`,并且在继承类中又定义了另外一种函数的实现.在运行时根据虚函数表(virtual table)来动态绑定具体哪种实现被执行.

纯虚函数是虚函数的一种,但纯虚函数在基类中没有被实现.如果类中存在纯虚函数,那么这个类(class)被称为抽象类(abstract class).抽象类不能被实例化,但能被继承.(译者注:当我们定义一个抽象类时,我们把继承类所共有的成员函数、变量写在抽象类中).抽象类的继承类必须定义所有的纯虚函数才能被实例化(译者注:否则仍然是抽象类).纯虚函数的写法是在虚函数的声明之后加上=0.

题目25. 为什么析构函数一般写成虚函数?构造函数可以写成虚函数吗?

参考答案: 析构函数一般推荐写成虚函数,这样能在有复杂继承关系时避免出错.因为构造函数在初始化的时候就需要确定所需构造对象的类型,然而虚函数是在运行时才确定具体形式, 所以构造函数不能是虚函数.

题目26. 编写一个函数:计算正整数的阶乘.

参考答案:

```
//使用For循环的方法实现
int factorial(int n){
    int output =1;
    for (int i =2 ; i <= n ; ++i)
        output *= i;
    return output;
}

//使用递归的方法实现
int factorial(int n){
   if (n == 0) return 1;
   return n*factorial(n-1);
}

//尾部递归的方法实现
int factorial(int n, int last = 1){
    if (n == 0) return last;
    return factorial(n-1, last * n);
}
```

题目27. 编写一个函数:输入一个数组,返回连续子数组的最大和.

参考答案:

```
#include <vector>
#include <algorithm> // std::max

using namespace std;

template <typename T>
T max_sub_array(vector<T> const & numbers){
    T max_ending = 0, max_so_far = 0;
    for(auto & number: numbers){
        max_ending = max(0, max_ending + number);
        max_so_far = max(max_so_far, max_ending);
    }
    return max_so_far;
}
```

题目28. 编写一个函数:输入一个正整数,输出它的所有质因子.

参考答案:

```
#include <vector>
using namepsace std;

vector<int> prime_factors(int n){
    vector<int> factors;
    for (int i = 2; i <= n/i ; ++i)
        while (n % i == 0) {
            factors.push_back(i);
            n /= i;
        }
    if (n > 1)
        factors.push_back(n);
    return factors;
}
```

题目29. 编写一个位操作函数:输入一个64位的二进制数,交换第i位与第j位.

参考答案:

```
long swap_bits (long x, const int &i, const int &j){
    if ( ((x >>i) & 1L) != ((x>>j) & 1L) )
        x ^= (1L <<i ) | (1L <<j);
    return x;
}
```

题目30. 编写一个函数:实现链表逆序(即“甲－＞乙－＞丙”变换成“丙－＞乙－＞甲”).

参考答案:

```
#include <memory> // shared_ptr
using namespace std;

template<typename T>
struct node_t {
    T data;
    shared_ptr<node_T<T>> next;
};

//递归的方式实现
template<typename T> shared_ptr<node_t<T>>
    reverse_linked_list(
        const shared_ptr<node_t <T>> &head){
    if (!head || !head->next) {
        return head;
    }

    shared_ptr<node_t<T>>
      new_head = reverse_linked_list(head->next);
    head->next->next = head;
    head->next = nullptr;
```

```
    return new_head;
}

//while循环的方式实现
template<typename T> shared_ptr<node_t<T>>
    reverse_linked_list(
        const shared_ptr<node_t <T>> &head){
    shared_ptr<node_t<T>>
        prev = nullptr, curr = head;
     while(curr) {
  shared_ptr<node_t<T>> temp = curr-> next;
  curr->next = prev;
  prev = curr;
  curr = temp;
     }
     return prev;
}
```

题目31. 编写一个函数:输入一个字符串(其中有左括号和右括号),输出括号是否成对.

参考答案:

```
#include<string>
#include<stack>
using namespace std;

bool is_par_balanced(const string input)
{
    //"()))())"=> false
    //"(a(dd)()(()))"=>true
    stack<char> par_stack;
    for(auto &c: input)
    {
        if(c==')')
        {
            if(par_stack.empty())
```

```
            return false;
            else if(par_stack.top()=='(')
            par_stack.pop();
        }
        else if(c=='(')
        par_stack.push(c);
    }
    return par_stack.empty();
}
```

题目32. 编写一个函数:输入一个二叉树(binary tree),返回这个二叉树的高度.

参考答案:

```
#include<memory> //std::shared_ptr
#include<algorithm> //std::max
using namespace std;

template <typename T>
struct BinaryTree {
    T data;
    shared_ptr<BinaryTree<T>> left, right;
};


template <typename T>
int height(
    const shared_ptr<BinaryTree<T>> &tree,
    int count = -1){
        if (!tree) return count;
        return max(
           height(tree->left, count + 1),
           height(tree->right, count +1));
}
```

5.5 蒙特卡洛模拟.数值方法.

题目1. 如何用蒙特卡洛计算π?计算结果的标准差为多少?

参考答案: 使用授拒测试(accept-reject testing)方法可以计算出π的大小.细节如下:均匀的在$[-1,1]\times[-1,1]$的正方形区域中生成N个点, 如果点(x,y) 的坐标落在内切圆$D(0,1)$之中(即(x,y)满足 $x^2+y^2\leq 1$)则接受,否则拒绝. 统计出被接受点的个数记做A.因为点是均匀产生的,面积之比等于概率之比.当$N\to\infty$时,接受概率$\frac{A}{N}$近似等于内切圆与正方形的面积之比.已知正方形面积为1,内切圆面积为$\frac{\pi}{4}$.所以

$$\pi \approx \frac{4A}{N}.$$

这种方法的标准误差为$O\left(\frac{1}{\sqrt{N}}\right)$. 为了让标准误差更加精确,下面计算$\frac{1}{\sqrt{N}}$的系数.结果请见 (5.79).

令 $U_1, U_2, \cdots$ 为独立同分布的二维随机变量(bivariate random variables).这些随机变量在 $[-1,1]\times[-1,1]$的区域里均匀分布. 记 $\mathbf{1}_{D(0,1)}$ 为区域 $D(0,1)$的指示函数(indication function), 可以写成

$$\mathbf{1}_{D(0,1)}(x,y) = \begin{cases} 1, & (x,y)\in D(0,1); \\ 0, & \text{其它}. \end{cases}$$

令

$$X_i = \mathbf{1}_{D(0,1)}(U_i), \ \forall\, i\geq 1.$$

则$X_i,\ i\geq 1$,为独立同分布的随机变量并且:

$$\begin{aligned} E[X_i] &= E\left[\mathbf{1}_{D(0,1)}(U_i)\right] = \iint_{D(0,1)} \frac{1}{4}dxdy \\ &= \frac{\pi}{4}. \end{aligned} \tag{5.77}$$

所以 $X_i,\ i\geq 1$的期望有界.根据强大数定理(strong law of large numbers) 有

$$\lim_{n\to\infty} \frac{X_1+X_2+\cdots+X_n}{n} = \frac{\pi}{4} \quad \text{两者以概率1收敛}.$$

$X_1 + X_2 + \cdots + X_n$ 代表有多少个点落在了 $D(0,1)$之中. 当 N 足够大时有:

$$\frac{X_1 + X_2 + \cdots + X_N}{N} \approx \frac{\pi}{4}. \qquad (5.78)$$

为了计算估计值(5.78)的标准误差, 当 $1 \leq i \leq N$时

$$\left(\mathbf{1}_{D(0,1)}(U_i)\right)^2 = \mathbf{1}_{D(0,1)}(U_i), \ \forall\, 1 \leq i \leq N,$$

使用(5.77)有:

$$\begin{aligned} \mathrm{var}(X_i) &= E[X_i^2] - (E[X_i])^2 \\ &= E\left[\left(\mathbf{1}_{D(0,1)}(U_i)\right)^2\right] - \left(\frac{\pi}{4}\right)^2 \\ &= E\left[\mathbf{1}_{D(0,1)}(U_i)\right] - \frac{\pi^2}{16} \\ &= \frac{\pi}{4} - \frac{\pi^2}{16} \\ &= \frac{\pi(4-\pi)}{16}. \end{aligned}$$

那么

$$\begin{aligned} & \mathrm{var}\left(\frac{X_1 + X_2 + \cdots + X_N}{N}\right) \\ = & \frac{1}{N^2}\left(\mathrm{var}(X_1) + \cdots + \mathrm{var}(X_N)\right) \\ = & \frac{1}{N^2} \cdot N\,\frac{\pi(4-\pi)}{16} \\ = & \frac{\pi(4-\pi)}{16N}. \end{aligned}$$

将标准误差的平方取根号,就得到了使用蒙特卡洛方法估计π的标准误差:

$$\frac{\sqrt{\pi(4-\pi)}}{4}\,\frac{1}{\sqrt{N}} \approx 0.82\,\frac{1}{\sqrt{N}}. \qquad (5.79)$$

题目2. 产生独立标准正态分布样本的方法有哪些?

参考答案: 最常见的一共有三种方法:

- Box–Muller方法 (使用 Marsaglia–Bray 算法,可以避免计算三角函数)
- 授拒测试方法
- 逆函数转换方法(Inverse Transform).

关于这三种方的具体细节,请参考Glasserman的著作 [2]. □

题目3. 如何用正态分布随机数生成器来生成符合几何布朗运动的股票价格走势?

参考答案: 假设一只股票的价格走势服从几何布朗运动那么它的价格时间序列满足如下性质:

$$dS_t = \mu S_t dt + \sigma S_t dW_t, \tag{5.80}$$

其中 μ 和 σ分别为几何布朗运动的漂移率和波动率, W_t 是一个维纳过程.将0到 T均匀分成m个离散时间间隔,每个间隔的时长为 $\delta t = \frac{T}{m}$,并且对于 $j = 0 : m$ 有$t_j = j\delta t$. 通过对(5.80) 在 t_j 到 t_{j+1}之间积分有:

$$S_{t_{j+1}} - S_{t_j} = \mu \int_{t_j}^{t_{j+1}} S_t dt + \sigma \int_{t_j}^{t_{j+1}} S_t dW_t. \tag{5.81}$$

使用下列近似:

$$\begin{aligned} \int_{t_j}^{t_{j+1}} S_t dt &\approx S_{t_j}(t_{j+1} - t_j) \\ &= S_{t_j}\delta t; \end{aligned} \tag{5.82}$$

$$\begin{aligned} \int_{t_j}^{t_{j+1}} S_t dW_t &\approx S_{t_j}(W_{t_{j+1}} - W_{t_j}) \\ &= S_{t_j}\sqrt{\delta t}\, Z_{j+1}, \end{aligned} \tag{5.83}$$

其中 Z_{j+1}为服从标准正态的随机变量, 因为 W_t 是维纳过程所以 $W_{t_{j+1}} - W_{t_j}$ 服从以0为均值$t_{j+1} - t_j = \delta t$为方差的

正态分布即

$$W_{t_{j+1}} - W_{t_j} = \sqrt{\delta t} Z_{j+1}. \tag{5.84}$$

并且对于$j = 1 : m$的所有 Z_j都是相互独立的.

使用随机数生成器生成一系列独立的标准正态分布样本记为z_1, z_2, ..., z_m .根据(5.81–5.83)描述的方法可以将(5.80)离散化成如下形式

$$S_{t_{j+1}} - S_{t_j} = \mu S_{t_j} \delta t + \sigma S_{t_j} \sqrt{\delta t} z_{j+1},$$

对于所有 $j = 0 : (m-1)$,可以进一步简化为

$$S_{t_{j+1}} = S_{t_j} \left(1 + \mu \delta t + \sigma \sqrt{\delta t} z_{j+1}\right),$$

.

分析上面的递归等式我们不难发现,这种离散化的最大缺点是序列有可能生成负数.但是股票的价格永远不可能为负数. 一条非负路径可以通过如下方法获得.使用伊藤公式(Itô's formula)求解(5.80):

$$d(\ln(S_t)) = \left(\mu - \frac{\sigma^2}{2}\right) dt + \sigma dW_t. \tag{5.85}$$

对 (5.85) 在区间t_j 到 t_{j+1}积分. 则有:

$$\begin{aligned} \ln(S_{t_{j+1}}) - \ln(S_{t_j}) &= \ln\left(\frac{S_{t_{j+1}}}{S_{t_j}}\right) \\ &= \left(\mu - \frac{\sigma^2}{2}\right)(t_{j+1} - t_j) + \sigma(W_{t_{j+1}} - W_{t_j}) \\ &= \left(\mu - \frac{\sigma^2}{2}\right) \delta t + \sigma \sqrt{\delta t} Z_{j+1}, \end{aligned} \tag{5.86}$$

其中 $j = 0 : (m-1)$.对(5.86)使用 (5.84)相同的方法进行离散化则有:

$$\ln\left(\frac{S_{t_{j+1}}}{S_{t_j}}\right) = \left(\mu - \frac{\sigma^2}{2}\right) \delta t + \sigma \sqrt{\delta t} z_{j+1},$$

对所有的 $j = 0 : (m-1)$,有如下形式:

$$S_{t_{j+1}} = S_{t_j} \exp\left(\left(\mu - \frac{\sigma^2}{2}\right)\delta t + \sigma\sqrt{\delta t} z_{j+1}\right).$$

这种方法生成的路径就不会产生负数. □

题目4. 如何用12个在$[0,1]$上的均匀分布的样本生成一个近似服从标准正态分布的样本?

参考答案: 如果记这12个相互独立的来自$[0,1]$的均匀分布样本分别为 $u_1, u_2, \ldots, u_{12}$那么

$$\sum_{i=1}^{12} u_i - 6 \tag{5.87}$$

服从标准正态分布.证明如下: 根据中心极限定理(central limit theorem)可知如果相互独立同分布的随机变量$X_i, i \geq 1$的期望 $E[X]$和标准差$\sigma(X)$不是无穷大,则有

$$\lim_{n\to\infty} \frac{\frac{1}{n}\left(\sum_{i=1}^n X_i\right) - E[X]}{\frac{\sigma(X)}{\sqrt{n}}} = Z, \tag{5.88}$$

其中Z 是标准正态分布.

令 $U_1, U_2, \ldots, U_{12}$为 12 个相互独立,抽样自$[0,1]$的均匀分布样本. 根据 (5.88) ,当 $n = 12$ 并且 $X_i = U_i$, $i = 1 : 12$, 有以下结论:

$$Z \approx \frac{\frac{1}{12}\left(\sum_{i=1}^{12} U_i\right) - E[U]}{\frac{\sigma(U)}{\sqrt{12}}}, \tag{5.89}$$

其中

$$\begin{aligned} E[U] &= \int_0^1 u\,du = \frac{1}{2}; \qquad (5.90)\\ \sigma^2(U) &= E[U^2] - (E[U])^2 = \int_0^1 u^2 du - \left(\frac{1}{2}\right)^2 \\ &= \frac{1}{3} - \frac{1}{4} = \frac{1}{12}, \end{aligned}$$

因此

$$\sigma(U) = \frac{1}{\sqrt{12}}. \tag{5.91}$$

根据(5.89–5.91)可知,

$$\begin{aligned} Z &\approx \frac{\frac{1}{12}\left(\sum_{i=1}^{12} U_i\right) - \frac{1}{2}}{\frac{1}{12}} \\ &= \sum_{i=1}^{12} U_i - 6, \end{aligned}$$

从而 (5.87)可以被用于近似的生成正态分布样本.

需要注意的是,这种方法生成的正态分布样本不是十分准确. 因为通过 (5.87)公式计算出的样本一定在 $[-6,6]$区间里,而真实的正态分布样本可以取到无穷大.所以建议同学们如果需要生成更加有效和准确的正态分布样本应当使用Box–Muller方法.这种方法只需要两个独立的服从均匀分布的随机数生成器,然后搭配使用授拒测试方法就可以很好的生成标准正态分布样本.

□

题目5. 蒙特卡洛的收敛速度是多少?

参考答案: 如果n是使用蒙特卡洛方法生成随机样本的数目,m为分割0 到 T的时间间隔数. 那么蒙特卡洛的收敛速度为

$$O\left(\max\left(\frac{T}{m}, \frac{1}{\sqrt{n}}\right)\right).$$

上述收敛速度的估计适用于多个标的物衍生产品的定价.注意收敛速率跟标的物的个数无关. 这是蒙特卡洛方法相较于有限微分和数值积分方法的最大优点(finite differences and numerical integration methods).

□

题目6. 方差缩减(variance reduction)有哪些"奇技淫巧"?

参考答案: 方差缩减顾名思义是对使用蒙特卡洛方法生成的样本估计与真实值的误差进行缩减.可以使估计值在抽样次

数相同的情况下更接近真实值,从而提高蒙特卡洛方法的效率. 这里介绍三种最常见的方法:

- 控制方差法(Control Variance)
- 对偶变量法(Antithetic Variables);
- 矩匹配(Moment Matching).

关于上述三种方法的细节和具体应用请参考 Glasserman的著作 [2]. □

题目7. 如何生成两个相关系数为ρ的正态分布样本?

参考答案: 设两个独立的标准正态分布样本分别为z_1 和 z_2,它们分别来自两个独立的标准正态分布记为Z_1 和 Z_2. 令$X_1 = Z_1$, $X_2 = \rho Z_1 + \sqrt{1-\rho^2} Z_2$.因为$X_2$为两个独立正态分布随机变量的线性组合,所以它也是一个服从正态分布的随机变量.

并且有:

$$\begin{aligned}
\operatorname{corr}(X_1, X_2) &= \operatorname{corr}(Z_1, \rho Z_1 + \sqrt{1-\rho^2} Z_2) \\
&= \rho \operatorname{corr}(Z_1, Z_1) \\
&\quad + \sqrt{1-\rho^2} \operatorname{corr}(Z_1, Z_2) \\
&= \rho \operatorname{var}(Z_1) \\
&= \rho,
\end{aligned}$$

根据定义 $\operatorname{var}(Z_1) = 1$, 且 Z_1 和 Z_2相互独立, $\operatorname{corr}(Z_1, Z_2) = 0$. 所以$X_1$ 与 X_2都服从正态分布并且之间的相关系数为ρ. 综上所述,通过对两个独立的标准正态分布样本z_1 和 z_2进行如下的处理

$$x_1 = z_1 \quad 和 \quad x_2 = \rho z_1 + \sqrt{1-\rho^2} z_2$$

就可以得到两个相关系数为ρ的服从正态分布的样本. □

题目8. 牛顿法的收敛阶数是多少?

参考答案: 如果牛顿法收敛,那么它以二阶收敛. 根据牛顿法,记初始猜测为 x_0,需要递归求解目标方程$f(x) = 0$,其中

$f:\mathbb{R}\to\mathbb{R}$. 下一循环的估计值可以通过下列公式计算得到:

$$x_{k+1} = x_k - \frac{f(x_k)}{f'(x_k)},\ \forall\, k \geq 0. \tag{5.92}$$

牛顿法的误差以二阶收敛可以被定义为: 设x^* 为 $f(x)=0$的一个解,假如$f(x)$的二阶导$f''(x)$连续,而且$f'(x^*)\neq 0$.当x_0跟 x^* 十分接近时,一定存在正整数$M>0$ 和 n_M使

$$\frac{|x_{k+1}-x^*|}{|x_k-x^*|^2} < M,\ \forall\, k \geq n_M. \tag{5.93}$$

上述结论(5.93)的直观理解如下: 根据假设 $f(x^*)=0$,将解代入到(5.92)的递归式中可以得到

$$\begin{aligned} & x_{k+1}-x^* \\ = \ & x_k - x^* - \frac{f(x_k)-f(x^*)}{f'(x_k)} \\ = \ & \frac{f(x^*)-f(x_k)+(x_k-x^*)f'(x_k)}{f'(x_k)}. \end{aligned} \tag{5.94}$$

与 $f(x)$在点x_k处泰勒展开形式相同.如果$f''(x)$ 连续,则存在点c_k, x^* 和 x_k 使

$$\begin{aligned} f(x^*) &= f(x_k)+(x^*-x_k)f'(x_k) \\ &\quad + \frac{(x^*-x_k)^2}{2}f''(c_k). \end{aligned} \tag{5.95}$$

根据 (5.94) 和 (5.95), 有

$$x_{k+1}-x^* = (x^*-x_k)^2\,\frac{f''(c_k)}{2f'(x_k)},$$

得到如下结论:

$$\frac{|x_{k+1}-x^*|}{|x_k-x^*|^2} = \left|\frac{f''(c_k)}{2f'(x_k)}\right|. \tag{5.96}$$

如果$f''(x)$ 和 $f'(x)$都是连续的函数并且$f'(x^*) \neq 0$, x_k和 x^*也十分接近的情况下, $\left|\frac{f''(c_k)}{2f'(x_k)}\right|$无限接近于 $\left|\frac{f''(x^*)}{2f'(x^*)}\right| < \infty$.

当x_k 和 x^*十分接近时, $\left|\frac{f''(c_k)}{2f'(x_k)}\right|$ 也是一致有界的,从而(5.96) 可以写成 (5.93).

□

题目9. 哪一种有限差分法(finite difference method)与三叉树模型(trinomial trees)相对应?

参考答案: 正向欧拉(Forward Euler). 正向欧拉作为一种显式有限差分方法,可以用离散化的思想解BS偏微分方程(Black–Scholes PDE). 其本质是使用有限微分的方法,将标的物价格和时间构成一个"网"."网" 上每一个结点的期权价格可以用前一时刻三个相邻结点的期权价格进行线性组合得到.这跟三叉树模型计算期权价格的方法十分类似.

如果三叉树模型里的标的物价格使用的是对数价格(即上涨与下跌的比率都是相对于对数价格$\ln(S)$而言), 与此同时对常系数偏微分方程进行$x = \ln(S)$的替换,然后对其使用正向欧拉求解.那么这种方法与三叉树模型的求解过程就是一模一样的.

□

5.6 概率论. 随机分析.

题目1. 什么是指数分布? 指数分布的均值和方差是什么?

参考答案: 指数分布随机变量X的概率密度函数为

$$f(x) = \begin{cases} \alpha e^{-\alpha x}, & x \geq 0; \\ 0, & x < 0. \end{cases}$$

其中, 参数$\alpha > 0$.
X的期望值和方差分别为

$$E[X] = \frac{1}{\alpha}, \quad \mathrm{var}(X) = \frac{1}{\alpha^2}.$$

以上结论可由分部积分得到

$$\begin{aligned} \int_0^\infty x e^{-\alpha x} dx &= -\frac{x e^{-\alpha x}}{\alpha}\Big|_0^\infty + \frac{1}{\alpha}\int_0^\infty e^{-\alpha x} dx \\ &= 0 - \frac{e^{-\alpha x}}{\alpha^2}\Big|_0^\infty = \frac{1}{\alpha^2}; \\ \int_0^\infty x^2 e^{-\alpha x} dx &= -\frac{x^2 e^{-\alpha x}}{\alpha}\Big|_0^\infty + \frac{2}{\alpha}\int_0^\infty x e^{-\alpha x} dx \\ &= 0 + \frac{2}{\alpha} \times \frac{1}{\alpha^2} = \frac{2}{\alpha^3}. \end{aligned}$$

则

$$E[X] = \int_{-\infty}^{\infty} x f(x)\, dx = \alpha \int_0^\infty x e^{-\alpha x}\, dx = \frac{1}{\alpha};$$

$$E[X^2] = \int_{-\infty}^{\infty} x^2 f(x)\, dx = \alpha \int_0^\infty x^2 e^{-\alpha x}\, dx = \frac{2}{\alpha^2}.$$

因此

$$\mathrm{var}(X) = E[X^2] - (E[X])^2 = \frac{2}{\alpha^2} - \left(\frac{1}{\alpha}\right)^2 = \frac{1}{\alpha^2}. \quad \square$$

题目2. X和Y是独立的指数分布随机变量, 期望分别为6和8, 求Y大于X的概率.

参考答案: X和Y的概率密度函数分别是

$$f_X(x) = \begin{cases} \frac{1}{6}e^{-\frac{x}{6}}, & x \geq 0; \\ \\ 0, & x < 0; \end{cases}$$

$$f_Y(y) = \begin{cases} \frac{1}{8}e^{-\frac{y}{8}}, & y \geq 0; \\ \\ 0, & y < 0. \end{cases}$$

由于X和Y相互独立, 它们的联合概率密度函数$f_{XY}(x,y)$是边缘概率密度函数的乘积, 即

$$\begin{aligned} f_{XY}(x,y) &= f_X(x)\, f_Y(y) \\ &= \begin{cases} \frac{1}{48}e^{-\frac{x}{6}-\frac{y}{8}}, & x \geq 0, y \geq 0; \\ \\ 0, & \text{其他}. \end{cases} \end{aligned}$$

令

$$A = \{(x,y) \in \mathbb{R}^2 : y \geq x\}.$$

Y大于X的概率可以通过计算如下双重积分得到

$$\begin{aligned} P(Y \geq X) &= \iint_A f_{XY}(x,y)dxdy \\ &= \frac{1}{48}\int_0^\infty \int_0^y e^{-\frac{x}{6}-\frac{y}{8}}\,dxdy \end{aligned}$$

(接上页)

$$\begin{aligned}
P(Y \geq X) &= \frac{1}{48}\int_0^{\infty} e^{-\frac{y}{8}} \left(\int_0^{y} e^{-\frac{x}{6}}\, dx\right) dy \\
&= \frac{1}{48}\int_0^{\infty} e^{-\frac{y}{8}} \left(-6e^{-\frac{x}{6}}\Big|_0^y\right) dy \\
&= \frac{1}{8}\int_0^{\infty} e^{-\frac{y}{8}} \left(1 - e^{-\frac{y}{6}}\right) dy \\
&= \frac{1}{8}\int_0^{\infty} \left(e^{-\frac{y}{8}} - e^{-\frac{7}{24}y}\right) dy \\
&= \frac{1}{8}\left(8 - \frac{24}{7}\right) \\
&= \frac{4}{7}. \quad \square
\end{aligned}$$

题目3. 泊松分布的期望值和方差是什么?

参考答案: 泊松分布随机变量X取非负整数 k的概率为

$$P(X = k) = \frac{e^{-\lambda}\lambda^k}{k!}, \ \forall\, k \geq 0,$$

其中, 参数$\lambda > 0$.
接下来我们证明X的期望值和方差为

$$E[X] = \lambda, \quad \mathrm{var}(X) = \lambda.$$

根据定义,

$$\begin{aligned}
E[X] &= \sum_{k=0}^{\infty} P(X = k) \cdot k = \sum_{k=1}^{\infty} \frac{e^{-\lambda}\lambda^k}{k!}\, k \\
&= e^{-\lambda}\lambda \sum_{k=1}^{\infty} \frac{\lambda^{k-1}}{(k-1)!}. \qquad (5.97)
\end{aligned}$$

e^t的泰勒级数展开是

$$e^t = \sum_{k=0}^{\infty} \frac{t^k}{k!},$$

那么

$$\sum_{k=1}^{\infty} \frac{\lambda^{k-1}}{(k-1)!} = e^{\lambda}; \tag{5.98}$$

$$\sum_{k=2}^{\infty} \frac{\lambda^{k-2}}{(k-2)!} = e^{\lambda}. \tag{5.99}$$

由(5.97)和(5.98), 易知$E[X] = \lambda$.
为了得到$\mathrm{var}(X)$, 我们先计算

$$\begin{aligned} E[X^2] &= \sum_{k=0}^{\infty} P(X=k)\cdot k^2 \\ &= \sum_{k=1}^{\infty} \frac{e^{-\lambda}\lambda^k}{k!}\, k^2 \\ &= e^{-\lambda}\sum_{k=1}^{\infty} \frac{k\lambda^k}{(k-1)!}, \end{aligned}$$

也可写作

$$\begin{aligned} E[X^2] &= e^{-\lambda}\sum_{k=1}^{\infty} \frac{(k-1)\lambda^k}{(k-1)!} + e^{-\lambda}\sum_{k=1}^{\infty} \frac{\lambda^k}{(k-1)!} \\ &= e^{-\lambda}\lambda^2\sum_{k=2}^{\infty} \frac{\lambda^{k-2}}{(k-2)!} + e^{-\lambda}\lambda\sum_{k=1}^{\infty} \frac{\lambda^{k-1}}{(k-1)!} \\ &= \lambda^2 + \lambda, \end{aligned}$$

最后一步用到了(5.98)和(5.99).
综上所述,

$$\mathrm{var}(X) = E[X^2] - (E[X])^2 = \lambda. \quad \square$$

题目4. 从单位圆内随机(等概率)选取一个点, 它到圆心的距离的期望是多少?

参考答案: 从单位圆内随机(等概率)选取的点到圆心的距离是$E\left[\sqrt{X^2+Y^2}\right]$,其中$(X,Y)$是单位圆$D$内的均匀分布. (X,Y)的概率密度函数是

$$f(x,y)=\begin{cases}\frac{1}{\pi}, & x\in D;\\ 0, & \text{其他}.\end{cases}$$

则

$$E\left[\sqrt{X^2+Y^2}\right] = \frac{1}{\pi}\iint_D \sqrt{x^2+y^2}\,dxdy. \tag{5.100}$$

转换到极坐标下 $x=r\cos\theta$, $y=r\sin\theta$, 其中$0\le r\le 1$, $0\le\theta<2\pi$, 并且 $dxdy=rdrd\theta$, 由式(5.100), 我们得到

$$\begin{aligned}
& E\left[\sqrt{X^2+Y^2}\right]\\
= & \frac{1}{\pi}\int_0^{2\pi}\int_0^1 \sqrt{r^2\left(\cos^2\theta+\sin^2\theta\right)}\,rdrd\theta\\
= & \frac{1}{\pi}\int_0^1\int_0^{2\pi} r^2 d\theta\,dr \qquad (5.101)\\
= & \frac{1}{\pi}\int_0^1 2\pi r^2\,dr = 2\int_0^1 r^2 dr\\
= & \frac{2}{3},
\end{aligned}$$

其中为了得到式(5.101)我们利用了$\cos^2\theta+\sin^2\theta=1,\forall\theta$.
□

题目5. 考虑两个联合正态分布的随机变量X和Y, 期望均为0, 方差均为1. 如果$\mathrm{cov}(X,Y)=\frac{1}{\sqrt{2}}$, 那么条件概率$P(X>0|Y<0)$是多少?

参考答案: 根据条件概率的定义,

$$P(X>0|Y<0)=\frac{P(X>0,Y<0)}{P(Y<0)}. \tag{5.102}$$

注意到由于Y是标准正态随机变量

$$P(Y < 0) \; = \; \frac{1}{2}. \tag{5.103}$$

为了计算$P(X > 0, Y < 0)$, 令

$$W \; = \; \sqrt{2}\,X - Y. \tag{5.104}$$

由于 $E[X] = E[Y] = 0$, $E[W] = 0$. 此外, 由于

$$\mathrm{var}(X) = \mathrm{var}(Y) = 1 \;\; , \;\; \mathrm{cov}(X, Y) = \frac{1}{\sqrt{2}},$$

我们得到

$$\begin{aligned}
\mathrm{var}(W) &= \mathrm{var}\left(\sqrt{2}\,X - Y\right) \\
&= \mathrm{var}\left(\sqrt{2}\,X\right) - 2\,\mathrm{cov}\left(\sqrt{2}X, Y\right) + \mathrm{var}(Y) \\
&= 2\mathrm{var}(X) - 2\sqrt{2}\,\mathrm{cov}(X, Y) + \mathrm{var}(Y) \\
&= 1,
\end{aligned}$$

以及

$$\begin{aligned}
\mathrm{cov}(W, Y) &= \mathrm{cov}\left(\sqrt{2}\,X - Y, Y\right) \\
&= \sqrt{2}\,\mathrm{cov}(X, Y) - \mathrm{var}(Y) \\
&= 0.
\end{aligned}$$

注意到由于X和Y符合联合正态分布, $W = \sqrt{2}\,X - Y$是一个正态随机变量. 再者, 因为 $E[W] = 0$, $\mathrm{var}(W) = 1$, $\mathrm{cov}(W, Y) = 0$, 所以 W和Y是相互独立的标准正态随机变量.

由式(5.104), 我们得到

$$X \; = \; \frac{1}{\sqrt{2}}(W + Y).$$

则事件$\{X>0,Y<0\}$的概率是

$$\begin{aligned} & P(X>0,Y<0) \\ = & P\left(\frac{1}{\sqrt{2}}(W+Y)>0,Y<0\right) \\ = & P(W+Y>0,Y<0). \end{aligned} \tag{5.105}$$

两条直线$w+y=0$和$y=0$ 把平面(w,y)分为四个楔形区域

$$\begin{aligned} R_1 &= \{w+y>0,y<0\}; \\ R_2 &= \{w+y>0,y>0\}; \\ R_3 &= \{w+y<0,y<0\}; \\ R_4 &= \{w+y<0,y>0\}. \end{aligned}$$

注意到

$$P(W+Y>0,Y<0) = P((W,Y)\in R_1). \tag{5.106}$$

由于W和Y是独立的正态随机变量, 它们的联合概率密度函数具有旋转不变性, 因此

$$P((W,Y)\in R_1) = P((W,Y)\in R_4); \tag{5.107}$$

$$P((W,Y)\in R_2) = P((W,Y)\in R_3); \tag{5.108}$$

$$P((W,Y)\in R_2) = 3P((W,Y)\in R_1);$$

如图5.1.

另外, 注意到由于$P(W+Y=0 \text{ or } Y=0)=0$,

$$\sum_{i=1}^{4} P((W,Y)\in R_i)=1, \tag{5.109}$$

由式(5.107–5.109)可得

$$\begin{aligned} 1 &= \sum_{i=1}^{4} P((W,Y)\in R_i) \\ &= 2P((W,Y)\in R_1)+2P((W,Y)\in R_2) \\ &= 8P((W,Y)\in R_1), \end{aligned}$$

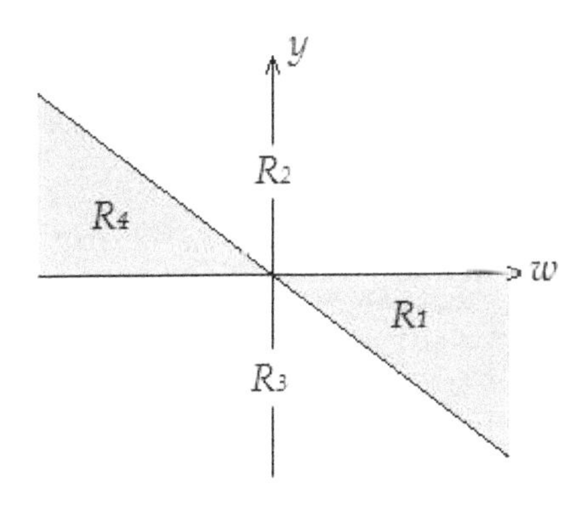

图 5.1: (w, y)平面上的区域R_1到R_4.

因此

$$P((W, Y) \in R_1) = \frac{1}{8}.$$

则

$$P(W + Y > 0, Y < 0) = \frac{1}{8}; \tag{5.110}$$

见式(5.106).

由式(5.105)及(5.110)可知

$$P(X > 0, Y < 0) = \frac{1}{8}. \tag{5.111}$$

由式(5.102), (5.103)及(5.111), 我们最终得到

$$\begin{aligned} P(X > 0|Y < 0) &= \frac{P(X > 0, Y < 0)}{P(Y < 0)} \\ &= \frac{1}{4}. \quad \square \end{aligned}$$

题目6. 如果X和Y均为对数正态随机变量, 它们的乘积也是对数正态分布吗?

参考答案: 首先, 易知如果X和Y是独立的对数正态随机变量, 那么XY也是对数正态随机变量, 由于 $\ln(XY) = \ln(X) + \ln(Y)$ 是两个独立的正态分布之和, 故而也是正态分布.

以上推断中独立性的假设可以进一步被放松.
如果$\ln(X)$和$\ln(Y)$遵从联合正态分布, 则$\ln(X)+\ln(Y)$也是正态分布, 故而XY遵从对数正态分布.

如果$\ln(X)$和$\ln(Y)$均为正态分布, 但不遵从联合正态分, 那么 $\ln(X)+\ln(Y)$不一定遵从正态分布, 故而XY也不一定是对数正态分布. □

题目7. X是期望为μ, 方差为σ^2的正态随机变量, Φ是标准正态分布的累积分布函数. 求$Y=\Phi(X)$的期望值.

参考答案: 令Z为独立于X的标准正态随机变量, 则

$$Y = \Phi(X) = P(Z \le X | X) = E[\mathbf{1}_{Z \le X} | X],$$

从而

$$E[Y] = E[E[\mathbf{1}_{Z\le X} | X]]. \qquad (5.112)$$

根据条件期望的重期望公式(Tower property)[10], 对于任意的随机变量T和W,

$$E[T] = E[E[T|W]]. \qquad (5.113)$$

令式(5.113) 中$T=\mathbf{1}_{Z\le X}$和$W=X$, 我们得到

$$\begin{aligned} E[E[\mathbf{1}_{Z\le X}|X]] &= E[\mathbf{1}_{Z\le X}] \\ &= P(Z\le X). \end{aligned} \qquad (5.114)$$

由式(5.112)和(5.114), 我们得到

$$E[Y] = P(Z\le X). \qquad (5.115)$$

注意到X是正态随机变量, 期望为μ, 方差 σ^2, 且X和Z是相互独立的.则 $Z-X$是正态随机变量, 其均值和方差分别

[10]为了计算T的期望值, 我们可以先计算已知W时T的条件期望, 然后对不同的W求平均.

为:

$$\begin{aligned} E[Z-X] &= -\mu; \\ \text{var}(Z-X) &= \text{var}(Z)+\text{var}(X) = 1+\sigma^2. \end{aligned}$$

所以$Z-X$是期望为$-\mu$, 方差为$1+\sigma^2$的正态随机变量. 因此, $\frac{Z-X+\mu}{\sqrt{1+\sigma^2}}$是标准正态随机变量, 我们由此得到

$$\begin{aligned} P(Z \le X) &= P(Z-X \le 0) \\ &= P\left(\frac{Z-X+\mu}{\sqrt{1+\sigma^2}} \le \frac{\mu}{\sqrt{1+\sigma^2}}\right) \\ &= \Phi\left(\frac{\mu}{\sqrt{1+\sigma^2}}\right). \end{aligned} \tag{5.116}$$

由式(5.115) 和(5.116)我们最终得到

$$E[Y] = \Phi\left(\frac{\mu}{\sqrt{1+\sigma^2}}\right). \tag{5.117}$$

我们注意到, 如果X是标准正态随机变量, 则$Y=\Phi(X)$是区间$[0,1]$上的均匀分布, 因此$E[Y]=\frac{1}{2}$. 在式(5.117)中当$\mu=0$和$\sigma=1$, 即X是标准正态随机变量时, $E[Y]=\Phi(0)=\frac{1}{2}$, 和上面的简单计算是相符的. □

题目8. 什么是大数定律?

参考答案: 大数定律有强大数定律和弱大数定律之分. 强大数定律说的是可积的独立同分布随机变量样本均值几乎处处收敛于分布的均值; 弱大数定律则只能保证依概率收敛.

准确来说, 令$X_1, X_2, \ldots$为一列独立同分布的随机变量, 它们存在有限的期望值 $\mu=E[X_i]$. 令$S_n = X_1+\cdots+X_n$.

强大数定律保证了几乎处处有 $\frac{S_n}{n} \to \mu$, 即

$$P\left(\lim_{n\to\infty}\frac{S_n}{n}=\mu\right) = 1. \tag{5.118}$$

弱大数定律保证了依概率有 $\frac{S_n}{n} \to \mu$, 即

$$\lim_{n\to\infty} P\left(\left|\frac{S_n}{n} - \mu\right| > \epsilon\right) = 0, \quad \forall\, \epsilon > 0. \tag{5.119}$$

注意到, 如果一列随机变量几乎处处收敛, 则它一定也依概率收敛, 因此如果式(5.118)成立, 则(5.119)也成立.从这种意义上来讲, 强大数定律"强于"弱大数定律. □

题目9. 什么是中心极限定理?

参考答案: 中心极限定理说独立同分布的(方差有限)随机变量序列之规范和(centered and scaled sum)的分布渐进于正态分布.

具体地说, 令X_1, X_2, ...为一个期望为$\mu = E[X_i]$, 方差为$\sigma^2 = \mathrm{var}[X_i]$的独立同分布随机变量序列.

定义$S_n = X_1 + \cdots + X_n$. 则

$$\lim_{n\to\infty} \frac{S_n - n\mu}{\sigma\sqrt{n}} = Z,$$

其中Z是标准正态分布, 上式中的收敛为依分布收敛, 也就是

$$\lim_{n\to\infty} P\left(\frac{S_n - n\mu}{\sigma\sqrt{n}} \le t\right) = P(Z \le t).$$

根据大数定律和中心极限定理, 如下关系在n趋于无穷大时的近似成立:

$$\frac{S_n}{n} \approx \mu + \frac{\sigma}{\sqrt{n}} Z. \quad \square$$

题目10. 什么是鞅(Martingale)?鞅和期权定价有什么联系?

参考答案: 令$(\Omega, \mathcal{F}_t, P)$为带滤子(filtration)的概率空间, 其中Ω是样本空间, $\mathcal{F}_t$是滤子, P是Ω上的概率测度.

随机过程X_t被称作滤子$\{\mathcal{F}_t\}$下的鞅当且仅当

(i) X_t是适应的, 即对于任意t, X_t是$\mathcal{F}_t$可测的;

(ii) X_t对于任意t可积, 即$E[|X_t|] < \infty$, $\forall t$;

(iii) 对于任意$s < t$, $E[X_t|\mathcal{F}_s] = X_s$ (几乎处处地).

换句话说, 如果给定时间s时所有可获得的信息$\mathcal{F}_s$, 鞅是一个随机过程, 其对未来时间t时随机过程取值的最优(最小二乘)估计, 即$E[X_t|\mathcal{F}_s]$, 也就是随机过程现在的取值X_s(几乎必然地).

鞅这个概念是期权定价理论的一块基石. 资产定价基本定理可以表述为只要市场是无套利的, 就存在风险中性概率使得资产的折现价格过程是鞅. 因此, 给衍生品定价的一种方法是运用伊藤公式推导出一个偏微分方程(称为定价方程)使得衍生品的折现价格过程在风险中性概率下是鞅.

题目11. 请解释用在伊藤引理推导过程中的假设$(dW_t)^2 = dt$.

参考答案: 微分形式的表述$(dW_t)^2 = dt$是传统积分表述(如下的黎曼积分)的简写,

$$\int_0^T (dW_t)^2 = \int_0^T dt. \qquad (5.120)$$

式(5.120)的直观解释涉及到布朗运动的二次变差(*quadratic variation*). 根据定义, 布朗运动W在区间$[0, T]$上的二次变差$QV_W[0, T]$定义为

$$QV_W[0,T] \ = \ \lim_{n\to\infty} \sum_{i=1}^{n} \left| W_{t_i} - W_{t_{i-1}} \right|^2,$$

其中 $t_i = \frac{i}{n}T$, $i = 0:n$. 因此我们有

$$\begin{aligned} & E\left[\lim_{n\to\infty}\sum_{i=1}^{n}\left|W_{t_i}-W_{t_{i-1}}\right|^2\right] \\ = \quad & \lim_{n\to\infty}\sum_{i=1}^{n}E\left[\left|W_{t_i}-W_{t_{i-1}}\right|^2\right] \\ = \quad & \lim_{n\to\infty}\sum_{i=1}^{n}(t_i-t_{i-1}) \\ = \quad & T, \end{aligned}$$

以上推导用到了对$i = 1:n$, $W_{t_i} - W_{t_{i-1}} \sim N(0, t_i - t_{i-1})$. 事实上, 可以证明下式为$L^2$收敛,

$$\sum_{i=1}^{n}\left|W_{t_i}-W_{t_{i-1}}\right|^2 \longrightarrow T \text{ as } n\to\infty$$

即,

$$\lim_{n\to\infty} E\left[\left|\sum_{i=1}^{n}\left|W_{t_i}-W_{t_{i-1}}\right|^2 - T\right|^2\right] = 0.$$

参考在传统的黎曼积分中使用的符号, 我们可以将其写作

$$\begin{aligned} \int_0^T (dW_t)^2 \quad &= \quad \lim_{n\to\infty}\sum_{i=1}^{n}\left|W_{t_i}-W_{t_{i-1}}\right|^2 \;=\; T \\ &= \quad \int_0^T dt, \end{aligned}$$

简写成微分形式为

$$(dW_t)^2 = dt. \quad \square$$

题目12. W_t是维纳过程, 求$E[W_t W_s]$.

参考答案: 假设 $s \le t$, 可以将W_t写作

$$W_t = (W_t - W_s) + W_s$$

进而有

$$W_t W_s = (W_t - W_s)W_s + W_s^2. \tag{5.121}$$

由W_τ, $\tau \geq 0$是维纳过程可知, $W_t - W_s$和W_s为相互独立的正态随机变量(均值为0), 因此

$$\begin{aligned} E[(W_t - W_s)W_s] &= E[W_t - W_s]E[W_s] \\ &= 0. \end{aligned} \tag{5.122}$$

另外, 因为W_s是正态分布, 均值为$E[W_s] = 0$, 方差为$\mathrm{var}(W_s) = s$, 我们有

$$\begin{aligned} \mathrm{var}(W_s) &= E[W_s^2] - (E[W_s])^2 \\ &= E[W_s^2], \end{aligned}$$

所以,

$$E[W_s^2] = \mathrm{var}(W_s) = s. \tag{5.123}$$

由式(5.121–5.123)可得

$$\begin{aligned} E[W_t W_s] &= E[(W_t - W_s)W_s] + E[W_s^2] \\ &= s. \end{aligned} \tag{5.124}$$

式(5.124)是在$s \leq t$的假设下推导出来的, 推广到一般情况, 我们得到

$$E[W_t W_s] = \min(s, t). \quad \square \tag{5.125}$$

题目13. W_t是维纳过程, 求$\mathrm{var}(W_t + W_s)$.

参考答案: 假设 $s \leq t$, 可以将W_t写作

$$W_t = (W_t - W_s) + W_s$$

进而有

$$W_t + W_s = (W_t - W_s) + 2W_s$$

则

$$\begin{aligned} \mathrm{var}(W_t + W_s) &= \mathrm{var}(W_t - W_s) \\ &\quad + 4\mathrm{cov}(W_t - W_s, W_s) \\ &\quad + 4\mathrm{var}(W_s). \end{aligned} \tag{5.126}$$

由W_τ, $\tau \geq 0$是维纳过程可知,

$$\mathrm{cov}(W_t - W_s, W_s) = 0; \tag{5.127}$$

$$\mathrm{var}(W_t - W_s) = t - s; \tag{5.128}$$

$$\mathrm{var}(W_s) = s, \tag{5.129}$$

由于 $W_t - W_s$和W_s是独立的正态分布变量, 方差分别为$t - s$和s.

由式(5.126–5.129), 我们得到

$$\mathrm{var}(W_t + W_s) = (t - s) + 4s = t + 3s. \tag{5.130}$$

由于式(5.130)是在$s \leq t$的假设下推导出来的, 我们得到

$$\mathrm{var}(W_t + W_s) = \max(s, t) + 3\min(s, t). \quad \square$$

题目14. W_t是一个维纳过程, 求

$$\int_0^t W_s\, dW_s \ \text{和}\ E\left[\int_0^t W_s\, dW_s\right].$$

参考答案: 由于 $\int x\, dx = \frac{x^2}{2} + C$, 我们先计算$d\left(\frac{W_t^2}{2}\right)$. 由伊藤引理, 如果 $f(x, t)$是连续可导函数, 则

$$df = \frac{\partial f}{\partial t}\, dt + \frac{\partial f}{\partial x}\, dW_t + \frac{1}{2}\frac{\partial^2 f}{\partial x^2}\, dt.$$

对于$f(x, t) = \frac{x^2}{2}$, 我们发现

$$d\left(\frac{W_t^2}{2}\right) = W_t dW_t + \frac{1}{2}dt,$$

从而

$$d\left(\frac{W_t^2 - t}{2}\right) = W_t dW_t. \tag{5.131}$$

通过从0到t积分式(5.131), 以及由于 $W_0 = 0$, 我们得到

$$\int_0^t W_s \, dW_s = \frac{W_t^2 - t}{2}, \ \forall \, t \geq 0. \tag{5.132}$$

由式(5.132), 可知

$$E\left[\int_0^t W_s \, dW_s\right] = \frac{E[W_t^2] - t}{2} = 0,$$

由于 W_t 是正态分布, 期望为0, 方差为t, 因此

$$\begin{aligned} E[W_t^2] &= \mathrm{var}(W_t) + (E[W_t])^2 \\ &= t. \quad \square \end{aligned}$$

题目15. 计算随机变量的概率密度分布

$$X = \int_0^1 W_t dW_t.$$

参考答案: 由伊藤引理, 如果W_t是维纳过程, $f(x)$的二阶导数连续, 则

$$df(W_t) = f'(W_t)dW_t + \frac{1}{2}f''(W_t)dt. \tag{5.133}$$

对于$f(x) = x^2$, 由式(5.133), 我们得到

$$dW_t^2 = 2W_t dW_t + dt. \tag{5.134}$$

通过从0到1对式(5.134)积分, 我们得到

$$W_1^2 = 2\int_0^1 W_t dW_t + 1 = 2X + 1. \tag{5.135}$$

注意到W_1是标准正态随机变量, 令$W_1 = Z$. 从式(5.135)解出X, 我们得到

$$X = \frac{W_1^2 - 1}{2} = \frac{Z^2 - 1}{2}.$$

令$f_X(x)$和$F_X(x)$ 分别为X的概率密度函数和累积分布函数.

注意到

$$P\left(X \leq -\frac{1}{2}\right) = P(Z^2 \leq 0) = 0.$$

因此, 如果$x < -\frac{1}{2}$, 则$F_X(x) = P(X \leq x) = 0$, 从而

$$f_X(x) = 0,$$

由于 $f_X(x) = F_X'(x)$.

如果$x \geq -\frac{1}{2}$, 则

$$\begin{aligned} F_X(x) &= P(X \leq x) = P\left(\frac{Z^2 - 1}{2} \leq x\right) \\ &= P\left(Z^2 \leq 2x + 1\right) \\ &= P\left(-\sqrt{2x+1} \leq Z \leq \sqrt{2x+1}\right) \\ &= 2\,P\left(0 \leq Z \leq \sqrt{2x+1}\right) \\ &= \frac{2}{\sqrt{2\pi}} \int_0^{\sqrt{2x+1}} e^{-\frac{y^2}{2}}\, dy. \end{aligned}$$

通过对$F_X(x)$微分, 可知对于$x < -\frac{1}{2}$,

$$\begin{aligned} f_X(x) &= F_X'(x) \\ &= \sqrt{\frac{2}{\pi}} \frac{d}{dx} \left(\int_0^{\sqrt{2x+1}} e^{-\frac{y^2}{2}}\, dy \right) \\ &= \sqrt{\frac{2}{\pi}} e^{-\frac{(\sqrt{2x+1})^2}{2}} \left(\sqrt{2x+1}\right)' \\ &= \sqrt{\frac{2}{\pi}} \frac{e^{-\frac{2x+1}{2}}}{\sqrt{2x+1}}. \end{aligned}$$

我们最终得到

$$f_X(x) = \begin{cases} \sqrt{\frac{2}{\pi}} \frac{e^{-\frac{2x+1}{2}}}{\sqrt{2x+1}}, & \text{if } x > -\frac{1}{2}; \\ 0 & \text{if } x \le -\frac{1}{2}. \end{cases} \quad \square$$

题目16. W_t是维纳过程, 求如下随机积分的均值和方差.

$$\int_0^t W_s^2 dW_s.$$

参考答案: 令$\mathcal{B}_t$为时间区间$[0,t]$上的Borel σ-代数, $\mathcal{F}_t$为概率空间的滤子, 维纳过程W_t定义在概率空间里.我们将用到以下两个结论: [11]

Theorem 5.1. *(Martingality)*
f_s在$[0,T]$上循序可测且平方可积, 即, 对于任意$t \in [0,T]$, f_t是$\mathcal{B}_t \otimes \mathcal{F}_t$-可测的, 且有$E\left[\int_0^T |f_t|^2 dt\right] < \infty$. 则随机积分$\int_0^t f_s dW_s$是期望为0, 平方可积的鞅.特别地,

$$E\left[\int_0^t f_s dW_s\right] = 0, \quad \forall\, t \in [0,T]. \tag{5.136}$$

Theorem 5.2. *(Itô's isometry)*
f_t和g_t均为循序可测且平方可积的过程.则

$$E\left[\int_0^t f_s dW_s \int_0^t g_s dW_s\right] = \int_0^t E[f_s\, g_s] ds.$$

特别地, 当$f_s = g_s$时, 我们有

$$E\left[\left(\int_0^t f_s dW_s\right)^2\right] = \int_0^t E[f_s^2] ds. \tag{5.137}$$

[11]定理 5.1和定理 5.2的证明分别参见参考文献[1]第65页的定理2.8和第67页的定理3.1

在本题中, 我们需要计算$E[X]$和$\mathrm{var}(X)$, 其中

$$X = \int_0^t W_s^2 dW_s.$$

先检验被积函数W_s^2在$[0,t]$上是否循序可测且平方可积.

W_s^2循序可测因其为适应的并且是连续的.

因为$W_s \sim N(0,s)$, 所以$W_s = \sqrt{s}Z$, 其中Z为标准正态随机变量.因此

$$E[W_s^4] = E\left[\left(\sqrt{s}Z\right)^4\right] = s^2\, E[Z^4] = 3s^2, \qquad (5.138)$$

其中, 我们用到了标准正态分布的四阶矩等于3的结论, 即, $E[Z^4] = 3$. 由(5.138)可知

$$\int_0^t E[W_s^4]ds = \int_0^t 3s^2 ds = t^3 < \infty. \qquad (5.139)$$

换句话说, W_s^2在$[0,t]$上平方可积.

综上, 我们可以对$f_s = W_s^2$和$g_s = W_s^2$运用定理5.1和定理5.2.

对$f_s = W_s^2$运用式(5.136)可得

$$E[X] = E\left[\int_0^t W_s^2 dW_s\right] = 0. \qquad (5.140)$$

对$f_s = g_s = W_s^2$运用式(5.137)可得

$$\begin{aligned} E[X^2] &= E\left[\left(\int_0^t W_s^2 dW_s\right)^2\right] = \int_0^t E[W_s^4]ds \\ &= t^3. \end{aligned}$$

由于$E[X] = 0$, 我们得到

$$\mathrm{var}(X) = E[X^2] - (E[X])^2 = t^3. \quad \square$$

题目17. W_t是维纳过程, 求如下随机积分的方差.

$$X = \int_0^1 \sqrt{t} e^{\frac{W_t^2}{8}} dW_t.$$

参考答案: 我们用定理5.1和定理5.2来解答本题. 先检验被积函数$\sqrt{t} e^{\frac{W_t^2}{8}}$在$[0,1]$上是否循序可测且平方可积.

$\sqrt{t} e^{\frac{W_t^2}{8}}$循序可测因其为适应的并且是连续的.

因为$W_t \sim N(0,t)$, W_t的概率密度函数是 $\frac{1}{\sqrt{2\pi t}} e^{-\frac{x^2}{2t}}$, 所以

$$\begin{aligned} E\left[e^{\frac{W_t^2}{4}}\right] &= \frac{1}{\sqrt{2\pi t}} \int_{-\infty}^{\infty} e^{\frac{x^2}{4}} e^{-\frac{x^2}{2t}} dx \\ &= \frac{1}{\sqrt{2\pi t}} \int_{-\infty}^{\infty} e^{\left(\frac{1}{4} - \frac{1}{2t}\right) x^2} dx \\ &= \frac{1}{\sqrt{2\pi t}} \int_{-\infty}^{\infty} e^{-\frac{1}{2} \cdot \frac{2-t}{2t} x^2} dx \\ &= \frac{1}{\sqrt{2\pi t}} \sqrt{2\pi \cdot \frac{2t}{2-t}} \qquad (5.141) \\ &= \sqrt{\frac{2}{2-t}}, \qquad (5.142) \end{aligned}$$

其中, 式(5.141)用到了以下等式: 对于任意常数$A > 0$,

$$\int_{-\infty}^{\infty} e^{-\frac{Ax^2}{2}} dx = \sqrt{\frac{2\pi}{A}},$$

由式(5.142)可知

$$\begin{aligned} \int_0^1 E\left[t e^{\frac{W_t^2}{4}}\right] dt &= \int_0^1 t\, E\left[e^{\frac{W_t^2}{4}}\right] dt \\ &= \int_0^1 t \cdot \sqrt{\frac{2}{2-t}} dt \\ &= -\frac{2\sqrt{2}}{3} \left((t+4)\sqrt{2-t}\right)\Big|_0^1 \\ &= \frac{2}{3}\left(8 - 5\sqrt{2}\right). \qquad (5.143) \end{aligned}$$

因此 $\int_0^1 E\left[te^{\frac{W_t^2}{4}}\right]dt < \infty$, 运用定理5.1和定理5.2的条件得到了满足.

由式(5.136)可知

$$E[X] = E\left[\int_0^1 \sqrt{t}e^{\frac{W_t^2}{8}}dW_t\right] = 0. \tag{5.144}$$

根据式(5.137)和(5.143), 我们得到

$$\begin{aligned} E[X^2] &= E\left[\left(\int_0^1 \sqrt{t}e^{\frac{W_t^2}{8}}dW_t\right)^2\right] \\ &= \int_0^1 E\left[\left(\sqrt{t}e^{\frac{W_t^2}{8}}\right)^2\right]dt \\ &= \int_0^1 E\left[te^{\frac{W_t^2}{4}}\right]dt \\ &= \frac{2}{3}(8-5\sqrt{2}). \end{aligned} \tag{5.145}$$

由式(5.144)和式(5.145)可知,

$$\mathrm{var}(X) = E[X^2] - (E[X])^2 = \frac{2}{3}\left(8-5\sqrt{2}\right). \quad \square$$

题目18. W_t是维纳过程, 求$E\left[e^{W_t}\right]$.

参考答案:

解答1: 如果W_t是维纳过程, 令$Y = e^{W_t}$, 那么

$$\ln(Y) = W_t \cong \sqrt{t}Z, \tag{5.146}$$

其中Z是标准正态随机变量, 所以Y是对数正态随机变量.

我们知道, 对数正态随机变量$\ln(V) = \mu + \sigma Z$的期望可由下式给出

$$E[V] = e^{\mu+\frac{\sigma^2}{2}}. \tag{5.147}$$

代入$\mu = 0$和$\sigma = \sqrt{t}$可得,

$$E\left[e^{W_t}\right] = E[Y] = e^{\frac{t}{2}}.$$

解答2: 由$W_t \cong \sqrt{t}Z$可知

$$\begin{aligned} E\left[e^{W_t}\right] &= E\left[e^{\sqrt{t}Z}\right] = \frac{1}{\sqrt{2\pi}} \int_{-\infty}^{\infty} e^{\sqrt{t}x} e^{-\frac{x^2}{2}}\, dx \\ &= \frac{1}{\sqrt{2\pi}} \int_{-\infty}^{\infty} e^{\sqrt{t}x - \frac{x^2}{2}}\, dx. \qquad (5.148) \end{aligned}$$

由配方法可得

$$\sqrt{t}x - \frac{x^2}{2} = -\frac{1}{2}\left(x - \sqrt{t}\right)^2 + \frac{t}{2},$$

所以

$$e^{\sqrt{t}x - \frac{x^2}{2}} = e^{\frac{t}{2}} e^{-\frac{1}{2}\left(x-\sqrt{t}\right)^2}. \qquad (5.149)$$

由式(5.148)和式(5.149)可知

$$\begin{aligned} E\left[e^{W_t}\right] &= e^{\frac{t}{2}} \frac{1}{\sqrt{2\pi}} \int_{-\infty}^{\infty} e^{-\frac{1}{2}\left(x-\sqrt{t}\right)^2}\, dx \\ &= e^{\frac{t}{2}} \frac{1}{\sqrt{2\pi}} \int_{-\infty}^{\infty} e^{-\frac{y^2}{2}}\, dy \qquad (5.150) \\ &= e^{\frac{t}{2}}, \qquad (5.151) \end{aligned}$$

式(5.150)进行了变量替换$y = x - \sqrt{t}$, 式(5.151)用到了标准正态分布概率密度函数的性质

$$\frac{1}{\sqrt{2\pi}} \int_{-\infty}^{\infty} e^{-\frac{y^2}{2}}\, dy = 1.$$

□

题目19. W_t是维纳过程, 求如下随机积分的方差.

$$\int_0^t s\, dW_s.$$

参考答案: 我们知道, 对于任意确定性的平方可积函数$f: [a,b] \to \mathbb{R}$, 随机积分$\int_a^b f(s)dW_s$是正态分布, 其期望为0, 方差等于f的二阶范数的平方, 即,

$$\int_a^b f(s)dW_s \sim N\left(0, \int_a^b f^2(s)ds\right).$$

则

$$\int_0^t s\,dW_s \sim N\left(0, \int_0^t s^2 ds\right) = N\left(0, \frac{t^3}{3}\right),$$

因此

$$\mathrm{var}\left(\int_0^t s dW_s\right) = \frac{t^3}{3}. \quad \square$$

题目20. W_t是维纳过程, 令

$$X_t = \int_0^t W_\tau d\tau. \tag{5.152}$$

X_t的分布是怎样的?X_t是鞅吗?

参考答案: 本题的一种解法(使用分部积分)已在第1章中给出.这里我们提供另一种解法.

X_t不是鞅, 因为如果我们把式(5.152)写成微分形式

$$dX_t = W_t dt = W_t dt + 0\,dW_t,$$

我们可以把X_t看作一个只有漂移项W_t的扩散过程.

我们知道单参数高斯随机变量族的积分仍为高斯(正态)分布. 由于W_t是高斯随机变量族, 所以X_t满足正态分布.因此我们有,

$$E[X_t] = \int_0^t E[W_\tau]d\tau = 0.$$

所以

$$\mathrm{var}(X_t) = E[X_t^2] - (E[X_t])^2 = E[X_t^2]. \tag{5.153}$$

注意到

$$X_t^2 = \left(\int_0^t W_s ds\right)^2 = \int_0^t \int_0^t W_s W_u ds du, \qquad (5.154)$$

由式(5.125), 我们有

$$E[W_s W_u] = \min\{s, u\}, \ \forall\, s, u > 0; \qquad (5.155)$$

根据式(5.153–5.155), 我们得到

$$\begin{aligned}
\mathrm{var}(X_t) &= E[X_t^2] = \int_0^t \int_0^t E[W_s W_u] ds du \\
&= \int_0^t \int_0^t \min\{s, u\} ds du \\
&= \int_0^t \left(\int_0^u s ds + \int_u^t u ds\right) du \\
&= \int_0^t \left(\frac{u^2}{2} + u(t-u)\right) du \\
&= \int_0^t \left(ut - \frac{u^2}{2}\right) du \\
&= \frac{t^3}{2} - \frac{t^3}{6} \\
&= \frac{t^3}{3}.
\end{aligned}$$

所以, X_t是期望为0、方差为$\frac{t^3}{3}$的正态分布, 即,

$$X_t \sim N\left(0, \frac{t^3}{3}\right). \quad \square$$

题目21. 什么是伊藤过程?

参考答案: 伊藤过程指的是满足以下随机微分方程(SDE)的随机过程X_t,

$$dX_t = a(X_t, t)dt + b(X_t, t)dW_t, \qquad (5.156)$$

其中W_t是维纳过程.

dt项的系数$a(x,t)$是X_t的漂移率； dW_t项的系数$b(x,t)$是X_t的扩散系数.

根据定义, SDE(5.156)是以下随机积分方程的简写

$$X_t = X_0 + \int_0^t a(X_s, s)\, ds + \int_0^t b(X_s, s)\, dW_s.$$

另外我们注意到, 随机微分方程解的存在性和唯一性的充分条件是漂移率$a(x,t)$和扩散系数$b(x,t)$均为局部Lipschitz函数, 并且随x的增长速率至多线性. □

题目22. 什么是伊藤引理?

参考答案: 伊藤引理(也叫伊藤公式)指出了如果X_t是满足以下SDE的伊藤过程

$$dX_t = a(X_t,t)dt + b(X_t,t)dW_t,$$

那么对于任意函数$f(x,t)$, 如果f有关于x的连续二阶导、关于t的连续一阶导, 那么$f(X_t,t)$也是维纳过程W_t驱动的伊藤过程.并且$f(X_t,t)$的漂移项和扩散项由$f(x,t)$的泰勒展开关于t的一次项和关于x的一、二次项决定.

$$\begin{aligned} df(X_t,t) &= \frac{\partial f}{\partial t}\, dt + \frac{\partial f}{\partial x}\, dX_t + \frac{1}{2}\frac{\partial^2 f}{\partial x^2}\, (dX_t)^2 \\ &= \frac{\partial f}{\partial t}\, dt + \frac{\partial f}{\partial x}\, [a(X_t,t)dt + b(X_t,t)dW_t] \\ &\quad + \frac{1}{2}\frac{\partial^2 f}{\partial x^2}\, [a(X_t,t)dt + b(X_t,t)dW_t]^2 \\ &= \left[\frac{\partial f}{\partial t} + a(X_t,t)\frac{\partial f}{\partial x} + \frac{b^2(X_t,t)}{2}\frac{\partial^2 f}{\partial x^2}\right] dt \\ &\quad + b(X_t,t)\frac{\partial f}{\partial x}\, dW_t. \end{aligned} \tag{5.157}$$

注意到式(5.157)中我们用到了

$$\begin{aligned} (dX_t)^2 &= [a(X_t,t)dt + b(X_t,t)dW_t]^2 \\ &= b^2(X_t,t)dt \end{aligned}$$

由于$(dW_t)^2 = dt$ (见本章题目5.6), $(dt)^2 = 0$, $dW_t\,dt = 0$.
□

题目23. W_t是维纳过程, $X_t = W_t^2$是鞅吗?

参考答案:

在带滤子的概率空间$(\Omega, \mathcal{F}_t, P)$上定义的随机过程$M_t$是鞅当且仅当

(i) M_t是适应的, 即对于任意t, M_t是$\mathcal{F}_t$可测的;

(ii) M_t对于任意t可积, 即$E[|M_t|] < \infty$, $\forall t$;

(iii) 对于任意$s < t$, $E[M_t|\mathcal{F}_s] = M_s$ (几乎处处地).

我们检验X_t是否满足以上的三个条件.

我们知道维纳过程W_t的任意连续函数是适应的, 所以X_t是适应的, 满足条件(i).

因为W_t是均值为0、方差为t的正态随机变量, 所以$E[X_t] = E[W_t^2] = \mathrm{var}[W_t] = t < \infty$. 因此, 对于任意$t$, 随机变量$X_t$可积, 满足条件(ii).

注意到, 对于任意$s < t$,

$$\begin{aligned} & E\left[X_t|\mathcal{F}_s\right] = E\left[W_t^2|\mathcal{F}_s\right] \\ = & E\left[(W_t - W_s + W_s)^2|\mathcal{F}_s\right] \\ = & E\left[(W_t - W_s)^2|\mathcal{F}_s\right] + 2E\left[(W_t - W_s)W_s|\mathcal{F}_s\right] \\ & + E\left[W_s^2|\mathcal{F}_s\right]. \end{aligned} \tag{5.158}$$

我们知道维纳过程W_t有独立的增量, 即$W_t - W_s$独立于$\mathcal{F}_s$; 同时增量是平稳的,即$W_t - W_s \sim W_{t-s}$; $W_t - W_s$是均值为0、方差为$t-s$的正态随机变量, 即 $E[W_{t-s}] = 0$, $\mathrm{var}(W_{t-s}) = E[W_{t-s}^2] = t - s$.所以

$$E\left[(W_t - W_s)^2|\mathcal{F}_s\right] = E[W_{t-s}^2] = t - s, \tag{5.159}$$

且

$$\begin{aligned} E\left[(W_t - W_s)W_s|\mathcal{F}_s\right] &= W_s E\left[W_t - W_s|\mathcal{F}_s\right] \\ &= W_s E[W_{t-s}] \\ &= 0. \end{aligned} \tag{5.160}$$

由于W_s是$\mathcal{F}_s$可测的, 我们有

$$E[W_s^2|\mathcal{F}_s] = W_s^2. \tag{5.161}$$

由式(5.158–5.161)可得

$$\begin{aligned} & E\left[X_t|\mathcal{F}_s\right] \\ = \;& E\left[(W_t - W_s)^2|\mathcal{F}_s\right] + 2E\left[(W_t - W_s)W_s|\mathcal{F}_s\right] \\ & + E\left[W_s^2|\mathcal{F}_s\right] \\ = \;& t - s + W_s^2. \end{aligned}$$

因此,

$$E\left[X_t|\mathcal{F}_s\right] \;=\; t - s + W_s^2 \;\neq\; W_s^2 \;=\; X_s,$$

所以X_t不满足条件(iii), X_t不是鞅. □

题目24. W_t是维纳过程, $N_t = W_t^3 - 3tW_t$是鞅吗?

参考答案:

解答1:

在带滤子的概率空间$(\Omega, \mathcal{F}_t, P)$上定义的随机过程$M_t$是鞅当且仅当

(i) M_t是适应的, 即对于任意t, M_t是$\mathcal{F}_t$可测的;

(ii) M_t对于任意t可积, 即$E[|M_t|] < \infty$, $\forall t$;

(iii) 对于任意$s < t$, $E[M_t|\mathcal{F}_s] = M_s$ (几乎处处地).

我们检验N_t是否满足以上的三个条件.

我们知道维纳过程W_t的任意连续函数是适应的, 所以N_t是适应的, 满足条件(i).

$$E[N_t] \;=\; E[W_t^3] - 3tE[W_t] \;=\; 0 \;<\; \infty,$$

由于$W_t \sim N(0, t)$, 所以$E[W_t] = E[W_t^3] = 0$. 因此对于任意t, 随机变量N_t可积, 满足条件(ii).

注意到, 对于任意$s < t$,

$$
\begin{aligned}
& E\left[N_t|\mathcal{F}_s\right] \\
= \; & E\left[W_t^3 - 3tW_t|\mathcal{F}_s\right] \\
= \; & E\left[(W_t - W_s + W_s)^3|\mathcal{F}_s\right] - 3tE\left[W_t|\mathcal{F}_s\right] \\
= \; & E\left[(W_t - W_s)^3|\mathcal{F}_s\right] + 3E\left[(W_t - W_s)^2 W_s|\mathcal{F}_s\right] \\
& + 3E\left[(W_t - W_s)W_s^2|\mathcal{F}_s\right] + E\left[W_s^3|\mathcal{F}_s\right] \\
& - 3tE\left[W_t|\mathcal{F}_s\right].
\end{aligned}
$$

我们知道维纳过程W_t有独立的增量, 即$W_t - W_s$独立于$\mathcal{F}_s$; 同时增量是平稳的, 即$W_t - W_s \sim W_{t-s}$; 另外W_s是$\mathcal{F}_s$可测的.则

$$
\begin{aligned}
E\left[(W_t - W_s)^3|\mathcal{F}_s\right] &= E[W_{t-s}^3] \\
&= 0; & (5.162) \\
E\left[(W_t - W_s)^2 W_s|\mathcal{F}_s\right] &= W_s \, E[W_{t-s}^2] \\
&= (t-s)W_s; & (5.163) \\
E\left[(W_t - W_s)W_s^2|\mathcal{F}_s\right] &= W_s^2 \, E[W_{t-s}] \\
&= 0; & (5.164) \\
E\left[W_s^3|\mathcal{F}_s\right] &= W_s^3, & (5.165)
\end{aligned}
$$

因为$W_{t-s} \sim N(0, t-s)$, 所以

$$E[W_{t-s}] = E[W_{t-s}^3] = 0;$$

$$E[W_{t-s}^2] = \mathrm{var}[W_{t-s}] = t - s.$$

又因为W_t是鞅, 我们有

$$E[W_t|\mathcal{F}_s] = W_s. \qquad (5.166)$$

由式(5.162–5.166)可得

$$\begin{aligned}
& E\left[N_t|\mathcal{F}_s\right] \\
= \;& E\left[(W_t-W_s)^3|\mathcal{F}_s\right]+3E\left[(W_t-W_s)^2W_s|\mathcal{F}_s\right] \\
& +3E\left[(W_t-W_s)W_s^2|\mathcal{F}_s\right] + E\left[W_s^3|\mathcal{F}_s\right] \\
& -3tE\left[W_t|\mathcal{F}_s\right] \\
= \;& 0+3(t-s)W_s+0+W_s^3-3tW_s \\
= \;& W_s^3-3sW_s \\
= \;& N_s.
\end{aligned}$$

所以, N_t满足条件(iii).综上, N_t是鞅.

解答2: 对N_t运用伊藤公式得到

$$\begin{aligned}
dN_t &= d(W_t^3-3tW_t) \\
&= 3W_t^2dW_t+\frac{1}{2}\cdot 6W_tdt - 3(W_tdt+tdW_t) \\
&= 3(W_t^2-t)dW_t.
\end{aligned}$$

N_t的漂移率为零, 且可以写为如下随机积分

$$N_t=\int_0^t 3(W_s^2-s)dW_s.$$

又因为

$$\begin{aligned}
& \int_0^t E\left[\left(3(W_s^2-s)\right)^2\right]ds \\
= \;& 9\left[\int_0^t \left(E[W_s^4]-2sE[W_s^2]+s^2\right)ds\right] \\
= \;& 9\left[\int_0^t \left(3s^2-2s^2+s^2\right)ds\right] \qquad (5.167) \\
= \;& 6t^3<\infty;
\end{aligned}$$

式(5.167)用到了$W_s = \sqrt{s}Z$, 其中Z是一个标准正态随机变量, 并且

$$\begin{aligned} E[W_s^2] &= E\left[\left(\sqrt{s}Z\right)^2\right] = sE[Z^2] = s; \\ E[W_s^4] &= E\left[\left(\sqrt{s}Z\right)^4\right] = s^2 E[Z^4] = 3s^2. \end{aligned}$$

因此对$s \in [0,t]$, $3(W_s^2 - s)$平方可积.根据定理5.1, 我们知道N_t是鞅.

题目25. 什么是Girsanov定理?

参考答案: Girsanov定理提供了一种通过用Radon-Nikodym导数定义新概率测度来改变维纳过程漂移率的途径.准确地说, 在$0 \leq t \leq T$上定义一个带滤子的概率空间$(\Omega, \mathcal{F}_t, P)$, 令$W_t$为概率测度$P$下的维纳过程.

令h_t为使下式为鞅的循序可测过程,

$$\mathcal{E}_t(h) = \exp\left(\int_0^t h_s dW_s - \frac{1}{2}\int_0^t h_s^2 ds\right)$$

用如下Radon-Nikodym导数定义Ω上的新概率测度$\widetilde{P}$:

$$\frac{d\widetilde{P}}{dP} = \exp\left(\int_0^T h_t dW_t - \frac{1}{2}\int_0^T h_t^2 dt\right). \qquad (5.168)$$

则维纳过程W_t在新概率测度$\widetilde{P}$下的漂移率为h.同样地, 如果我们定义$\widetilde{W}_t = W_t - h_t$, 那么$\widetilde{W}_t$是$\widetilde{P}$测度下的维纳过程.一般地, 如果$X_t$是一个扩散过程, 满足概率测度$P$下的SDE

$$dX_t = a(X_t,t)dW_t + b(X_t,t)dt$$

那么在由式(5.168)定义的概率测度$\widetilde{P}$下, X_t满足如下SDE

$$\begin{aligned} dX_t &= a(X_t,t)dW_t + b(X_t,t)dt \\ &= a(X_t,t)\left(d\widetilde{W}_t + h_t dt\right) + b(X_t,t)dt \\ &= a(X_t,t)d\widetilde{W}_t + \left(b(X_t,t) + h_t a(X_t,t)\right)dt. \end{aligned}$$

换言之, 在P测度下, X_t的漂移率为b, 扩散系数为a; 在$\widetilde{P}$测度下, 扩散系数保持不变, 漂移率变为$b + h\,a$.特别地, 如果我们令h为$-\frac{b}{a}$, 则X_t在$\widetilde{P}$测度下的漂移率为0 (在给定随机指数$\mathcal{E}_t(h)$是鞅的情况下).

另外我们注意到随机指数$\mathcal{E}_t(h)$是鞅[12]保证了用式(5.168)定义的新测度$\widetilde{P}$是一个概率测度, 即$\int_\Omega d\widetilde{P} = 1$, 由于

$$\begin{aligned}
\int_\Omega d\widetilde{P} &= \int_\Omega \frac{d\widetilde{P}}{dP} dP \\
&= E\left[\frac{d\widetilde{P}}{dP}\right] \\
&= E\left[\mathcal{E}_T(h)\right] \\
&= \mathcal{E}_0(h) \\
&= 1. \quad \square
\end{aligned}$$

题目26. 什么是鞅表示定理?其与期权定价和对冲有什么关系?

参考答案: 令W_t为定义在带有滤子的概率空间$(\Omega, \mathcal{F}_t, P)$上的维纳过程, 其中非降的$\sigma$-代数流$\{\mathcal{F}_t\}$由维纳过程$W_t$生成. 令$M_t$为关于$\{\mathcal{F}_t\}$的鞅, 使得对于任意$t$, M_t平方可积, 即, $E[M_t^2] < \infty$, $\forall t$.

鞅表示定理给出, 对于任意满足以上定义的鞅M_t, 存在一个平方可积的$\mathcal{F}_t$-适应过程θ_t, 使得M_t具有如下随机积分表示(几乎处处地)

$$M_t = E[M_0] + \int_0^t \theta_s dW_s$$

而且注意到鞅M_t是连续的.

---

[12] 保证$\mathcal{E}_t(h)$是鞅的充分条件是Novikov条件

$$E\left[e^{\frac{1}{2}\int_0^T h_t^2 dt}\right] < \infty.$$

鞅表示定理和期权定价对冲的关系如下.简单起见，我们假设无风险利率为零.假设标的资产的价格S_t服从在风险中性概率P下定义的SDE决定的扩散过程

$$dS_t = \sigma_t S_t dW_t$$

其中维纳过程W_t定义在带有滤子的概率空间$(\Omega, \mathcal{F}_t, P)$上，非降的$\sigma$-代数流$\{\mathcal{F}_t\}$由$W_t$生成. 考虑一个衍生品，其在到期时$(T)$的收益函数(可能是路径依赖的)是$\varphi_T$.资产定价基本定理给出，由于无风险利率被假定为零，衍生品的价值过程V_t是风险中性概率下的鞅.事实上，

$$V_t = E[\varphi_T | \mathcal{F}_t].$$

同时，如果收益函数φ_T平方可积，即$E[\varphi_T^2] < \infty$，那么V_t是平方可积的鞅.因此根据鞅表示定理，我们知道存在$\mathcal{F}_t$-适应过程θ_t使得

$$\begin{aligned}\varphi_T &= V_T = E[V_0] + \int_0^T \theta_t dW_s \\ &= E[V_0] + \int_0^T \frac{\theta_t}{\sigma_t S_t} dS_t.\end{aligned}$$

注意到$\frac{\theta_t}{\sigma_t S_t}$就是为了动态对冲衍生品所应该持有的标的资产数量.特别地，如果φ_T是看涨期权的收益函数，则$\frac{\theta_t}{\sigma_t S_t}$对应着看涨期权的delta. □

题目27. 求解

$$dY_t = Y_t dW_t, \tag{5.169}$$

其中W_t是个维纳过程.

参考答案:

解答*1:* 注意到式(5.169)是如下随机微分方程在$\mu = 0$和$\sigma = 1$时的特例

$$dS_t = \mu S_t dt + \sigma S_t dW_t, \tag{5.170}$$

上式是Black–Scholes框架下标的资产价格演化的模型.
式(5.170)的解为

$$S_t = S_0 \exp\left(\left(\mu - \frac{\sigma^2}{2}\right) t + \sigma\sqrt{t}Z\right), \tag{5.171}$$

其中$t > 0$, Z是标准正态随机变量. 在式(5.171)中令$\mu = 0$和$\sigma = 1$, 我们得到式(5.169)的解为

$$Y_t = Y_0 \exp\left(-\frac{t}{2} + \sqrt{t}Z\right). \tag{5.172}$$

解答2: 由伊藤引理可知, 若Y_t是一个满足随机微分方程$dY_t = Y_t dW_t$的随机过程, $f(y)$有连续的二阶导, 则

$$d(f(Y_t)) = \frac{1}{2} f''(Y_t) Y_t^2 dt + f'(Y_t) Y_t dW_t. \tag{5.173}$$

令$f(y) = \ln(y)$可知

$$d(\ln(Y_t)) = -\frac{1}{2}dt + dW_t. \tag{5.174}$$

对式(5.174)做积分,

$$\ln(Y_t) - \ln(Y_0) = -\frac{t}{2} + W_t - W_0 = -\frac{t}{2} + W_t,$$

上式用到了$W_0 = 0$.我们因此得到

$$Y_t = Y_0 \exp\left(-\frac{t}{2} + W_t\right).$$

由于W_t是维纳过程, W_t是均值为0、方差为t的正态随机变量, 即$W_t = \sqrt{t}Z$, 其中Z是标准正态随机变量. 所以, Y_t可以写作

$$Y_t = Y_0 \exp\left(-\frac{t}{2} + \sqrt{t}Z\right),$$

与式(5.172)一致. □

题目28. 求解以下SDE:

(i) $dY_t = \mu Y_t dt + \sigma Y_t dW_t$;

(ii) $dX_t = \mu dt + (aX_t + b) dW_t$.

参考答案:
(i) 根据伊藤引理, 如果Y_t满足SDE

$$dY_t = \mu Y_t dt + \sigma Y_t dW_t,$$

并且$f(y)$有连续的二阶导, 那么

$$\begin{aligned} df(Y_t) &= f'(Y_t)dY_t + \frac{1}{2}f''(Y_t)\sigma^2 Y_t^2 dt \\ &= \left[\mu Y_t f'(Y_t) + \frac{\sigma^2}{2}Y_t^2 f''(Y_t)\right] dt \\ &\quad + \sigma Y_t f'(Y_t) dW_t. \end{aligned} \tag{5.175}$$

把$f(y) = \ln(y)$代入式(5.175), 我们得到

$$d\ln(Y_t) = \left(\mu - \frac{\sigma^2}{2}\right) dt + \sigma dW_t. \tag{5.176}$$

对式(5.176)做从0到t的积分, 我们得到

$$\ln(Y_t) - \ln(Y_0) = \left(\mu - \frac{\sigma^2}{2}\right) t + \sigma W_t, \tag{5.177}$$

从式(5.177)解出Y_t, 我们得到

$$Y_t = Y_0 e^{\left(\mu - \frac{\sigma^2}{2}\right)t + \sigma W_t}.$$

(ii) 为了求解SDE

$$dX_t = \mu dt + (aX_t + b) dW_t \tag{5.178}$$

我们尝试如下形式的解

$$X_t = U_t V_t, \tag{5.179}$$

U_t是以下SDE的解

$$dU_t = aU_t dW_t, \quad \text{with } U_0 = 1, \tag{5.180}$$

V_t是以下SDE的解

$$dV_t = \alpha_t dt + \beta_t dW_t, \quad \text{with } V_0 = X_0, \tag{5.181}$$

其中系数α_t和β_t待定.

由(i)可知SDE

$$dY_t = \mu Y_t dt + \sigma Y_t dW_t$$

的解为

$$Y_t = Y_0 \, e^{\left(\mu - \frac{\sigma^2}{2}\right)t + \sigma W_t}. \tag{5.182}$$

在式(5.182)中, 让$\mu = 0$和$\sigma = a$, 我们可以得到SDE(5.180)的解

$$U_t = e^{aW_t - \frac{a^2}{2}t}. \tag{5.183}$$

注意到

$$\begin{aligned} dU_t \, dV_t &= (aU_t dW_t)\,(\alpha_t dt + \beta_t dW_t) \\ &= a\beta_t U_t dt, \end{aligned} \tag{5.184}$$

以上推导中用到了$(dW_t)^2 = dt$和$dW_t dt = 0$ (因为这项"相当于"$(dt)^{3/2}$). 根据伊藤乘积法则以及式(5.184), 我们有

$$\begin{aligned} dX_t &= d(U_t V_t) \;=\; U_t dV_t + V_t dU_t + dU_t \, dV_t \\ &= U_t(\alpha_t dt + \beta_t dW_t) + V_t(aU_t dW_t) \\ &\quad + a\beta_t U_t dt \\ &= (a\beta_t + \alpha_t)U_t dt + (aU_t V_t + \beta_t U_t)dW_t \\ &= (a\beta_t + \alpha_t)U_t dt + (aX_t + \beta_t U_t)dW_t \end{aligned} \tag{5.185}$$

则$X_t = U_t V_t$是SDE(5.178)的解当且仅当式(5.178)和式(5.185)中dW_t、dt两项的系数相等.

从dW_t项的系数相等我们可以得到

$$\begin{aligned} & aX_t + b = aX_t + \beta_t U_t \\ \Longleftrightarrow \quad & \beta_t U_t = b \\ \Longleftrightarrow \quad & \beta_t = bU_t^{-1} = b\, e^{-aW_t + \frac{a^2}{2}t}. \end{aligned} \tag{5.186}$$

从dt项的系数相等我们可以得到

$$\begin{aligned} & (a\beta_t + \alpha_t)U_t = \mu \\ \Longleftrightarrow \quad & a\beta_t + \alpha_t = \mu U_t^{-1} \\ \Longleftrightarrow \quad & \alpha_t = \mu\, U_t^{-1} - a\beta_t \\ \Longleftrightarrow \quad & \alpha_t = \mu\, e^{-aW_t + \frac{a^2}{2}t} - abe^{-aW_t + \frac{a^2}{2}t} \\ \Longleftrightarrow \quad & \alpha_t = (\mu - ab)\, e^{-aW_t + \frac{a^2}{2}t}. \end{aligned} \tag{5.187}$$

注意到SDE(5.181)的解V_t可由下式给出

$$V_t = X_0 + \int_0^t \alpha_s ds + \int_0^t \beta_s dW_s. \tag{5.188}$$

由式(5.179), 式(5.183), 和式(5.188), 我们可以得到式(5.178)的解

$$X_t \;=\; e^{aW_t - \frac{a^2}{2}t} \left(X_0 + \int_0^t \alpha_s ds + \int_0^t \beta_s dW_s \right),$$

并且, 用式(5.186)和式(5.187), 我们得到

$$\begin{aligned} X_t \;&=\; e^{aW_t - \frac{a^2}{2}t} \left[X_0 + \int_0^t (\mu - ab)\, e^{-aW_s + \frac{a^2}{2}s} ds \right. \\ &\quad \left. + \int_0^t be^{-W_s + \frac{a^2}{2}s} dW_s \right] \\ &=\; X_0\, e^{aW_t - \frac{a^2}{2}t} \\ &\quad + (\mu - ab) \int_0^t e^{a(W_t - W_s) - \frac{a^2}{2}(t-s)} ds \\ &\quad + b \int_0^t e^{a(W_t - W_s) - \frac{a^2}{2}(t-s)} dW_s. \quad \square \end{aligned}$$

题目29. 什么是Heston模型?

参考答案: Heston随机波动率模型假设标的资产的价格变化满足与对数正态模型相同的随机微分方程, 即,

$$dS_t \;=\; \mu S_t dt \;+\; \sqrt{v_t} S_t dW_t,$$

其中瞬时方差v_t遵循均值回归的CIR(Cox-Ingersoll-Ross)过程

$$dv_t = \lambda(v_t - m)dt + \eta\sqrt{v_t}dZ_t,$$

其中 λ和m均为正常数.维纳过程W_t和Z_t之间的相关系数为ρ, 即,

$$\mathrm{corr}(dW_t, dZ_t) = \rho dt.$$

Heston模型是一个在衍生品定价中常用到的基准模型.
Heston模型有如下特征:

- Heston模型考虑了杠杆效应, 也就是说维纳过程W_t和Z_t是相关的; 实证上, 相关系数ρ小于0, 这也是我们称之为杠杆效应的原因;
- Heston模型有欧式期权价格的伪解析解(其中包含傅里叶逆变换), 这使得校准模型中的参数更加容易;
- Heston模型的方差遵循一个回归速率为λ、长期均值为m的均值回归的过程.

注意到, 当$2\lambda m < \eta^2$时, 波动率可能会取到0.金融从业者通常假设在波动率取到0时是吸收的(取到0之后一直为0), 或者反射的(取到0之后反弹回正数).另外, 波动率的另一个边界无穷大是无法取到的. □

5.7 脑筋急转弯.

题目1. 一只跳蚤从相距100英寸的点A跳到点B(始终朝着点B跳).每一跳它都可以选择跳一英寸或者两英寸.请问这只跳蚤有多少种不同的跳法?

参考答案: 用a_n表示在相距n英寸的两点之间不同跳法的数量.我们想求a_{100}.

由于跳蚤可以选择一次跳一英寸或者两英寸, 它的最后一跳有两种可能: 从第$n-1$英寸的末端跳一英寸到终点, 或者从第$n-2$英寸的末端跳两英寸到终点. 因此, 跳n英寸的跳法数量是跳$n-1$英寸的跳法数量与跳$n-2$英寸的跳法数量之和.

我们得到如下$n > 2$时的递归式,

$$a_n = a_{n-1} + a_{n-2}. \tag{5.189}$$

显然, 跳蚤跳1英寸只有一种跳法, $a_1 = 1$; 跳2英寸有两种跳法(跳两次一英寸, 或者跳一次两英寸), $a_2 = 2$.

注意到(5.189)是斐波那契数列的递推关系,

$$\phi_1 = \frac{1+\sqrt{5}}{2}, \quad \phi_2 = \frac{1-\sqrt{5}}{2}$$

是式(5.189)的特征方程$x^2 - x - 1 = 0$的根.

则

$$a_n = C_1\phi_1^n + C_2\phi_2^n, \ \forall\, n \ge 1,$$

其中常数C_1和C_2的取值使得$a_1 = 1$以及$a_2 = 2$.

通过解以下线性系统

$$\begin{cases} C_1\phi_1 + C_2\phi_2 & = & 1 \\ C_1\phi_1^2 + C_2\phi_2^2 & = & 2, \end{cases}$$

我们得到

$$\begin{aligned} C_1 &= \frac{1+\sqrt{5}}{2\sqrt{5}} = \frac{\phi_1}{\sqrt{5}}; \\ C_2 &= -\frac{\sqrt{5}-1}{2\sqrt{5}} = -\frac{\phi_2}{\sqrt{5}}, \end{aligned}$$

因此

$$a_n = \frac{1}{\sqrt{5}}\left(\phi_1^{n+1} - \phi_2^{n+1}\right).$$

把$n = 100$代入以上表达式可以得到本题答案. □

题目2. 一个袋子里有三张饼: 一张两边都烤得金黄, 第二张饼两面都烤焦了, 第三张饼一面金黄, 一面烤焦了.从袋中随机拿出一张饼, 发现其中一面是金黄的, 求另外一面也是金黄色的概率.

参考答案:

解答1: 依据烧焦的面的数量将三张饼标记为0, 1, 2. 令E_i, $i = 0, 1, 2$表示从袋中取出第i号饼的事件.令A表示我们看到的(随机取出的饼)面是金黄色的事件.

根据贝叶斯公式, 在事件A发生的条件下, 我们取出0号饼的条件概率是

$$\begin{aligned}
P(E_0|A) &= \frac{P(E_0 \cap A)}{P(A)} = \\
&\frac{P(E_0)P(A|E_0)}{P(A|E_0)P(E_0) + P(A|E_1)P(E_1) + P(A|E_2)P(E_2)} \\
&= \frac{\frac{1}{3}\cdot 1}{1\cdot\frac{1}{3} + \frac{1}{2}\cdot\frac{1}{3} + 0\cdot\frac{1}{3}} \\
&= \frac{2}{3}.
\end{aligned}$$

解答2: 在我们所能看到的六个面中, 有三个金黄色的面; 其中有两个面属于两面都是金黄色的那张饼.因此, 饼的另一面也是金黄色的概率是 $\frac{2}{3}$. □

题目3. 爱丽丝和鲍勃玩掷硬币的游戏, 爱丽丝扔了$n + 1$枚, 鲍勃扔了n枚.假设他们用的硬币都是公平的, 求爱丽丝扔出的正面数量严格多于鲍勃的概率.

参考答案:

解答1: 爱丽丝扔的硬币数目多于鲍勃，因此她扔的正面或者背面中至少有一种多于鲍勃.然而，她也不可能既比鲍勃扔更多的正面，也比他扔更多的背面，因为她只多扔了一枚.所以爱丽丝要么扔了更多的正面，要么扔出更多的背面，其中有且仅有一个事件会发生.

由于这两个事件发生的概率是对等的，因此它们发生的概率均为$\frac{1}{2}$，所以爱丽丝扔的正面严格多于鲍勃的概率是 $\frac{1}{2}$.

解答2: 假设爱丽丝和鲍勃先各扔了n枚硬币. 令p为爱丽丝比鲍勃得到正面数量多的概率，q为爱丽丝和鲍勃得到正面数量相同的概率. 注意到由于爱丽丝得到的正面数量比鲍勃多和比鲍勃少的概率相同，所以$2p+q=1$.

接下来，爱丽丝扔第$(n+1)$枚硬币.为了使爱丽丝得到更多数量的正面，她要么在扔最后一枚硬币前就比鲍勃正面多，要么在扔最后一枚硬币前正面数目和鲍勃相同，且第$(n+1)$枚硬币扔出正面. 因此，爱丽丝扔出更多正面的概率是$p+\frac{1}{2}q$.

由于$2p+q=1$, $p+\frac{1}{2}q=\frac{1}{2}$. □

题目4. 爱丽丝要在三种甜点中艰难地做出选择. 如果她只有一枚公平的硬币，且她对三种甜品的喜好程度相当，那么她将如何使用这枚硬币帮助她挑选一种甜品呢?

参考答案:

解答1: 将三种甜品记为A, B, C. 首先假定硬币是公平的.将正面记为H，背面记为T. 在三种甜品中等概率选择的方法如下：扔两次硬币；如果结果为TH, HT, 或TT, 则分别对应甜品A, B, C；否则重复这个过程.

注意这个方法不必重复的概率是 $p=\frac{3}{4}$；因此，重复次数是以p为参数的几何分布随机变量.根据几何分布的期望公式,整个过程需要的重复次数的期望是$\frac{1}{p}=\frac{4}{3}$.由于每次需要扔两次硬币，爱丽丝做出决定期望的扔硬币次数是$\frac{8}{3}$.

现在，假定硬币正面和背面的概率不相等.一种以等概率选择甜品的方法是：扔四次硬币；如果得到序列$THHT$, $HTTH$, 或$THTH$, 则分别对应甜品A, B, C；如果得到其它的序列则重复这个过程.

解答2: 另一种方法是: 扔三次硬币; 如果得到HTT, THT, 或TTH则分别选择甜品A, B, C; 否则重复这个过程.

用和公平硬币相似的计算方法, 我们得到, 当硬币正面和背面的概率不等的时候, 这两种方法分别需要扔 $\frac{64}{3}$和8次硬币. □

题目5. 扔一枚公平的硬币直到得到正面, 需要的次数的期望是多少?假如这枚硬币得到正面的概率是p,期望又为多少?

参考答案: 用X表示扔硬币直到扔出正面的概率.如果第一次就扔出正面(概率是$\frac{1}{2}$), 则$X = 1$. 如果第一次扔出的是背面(概率也是$\frac{1}{2}$), 则我们相当于重置了扔硬币的过程, 在这种情况下得到1次正面需要的次数的期望为原本的期望加上1. 换句话说, 扔硬币次数的期望$E[X]$ 满足如下方程

$$E[X] = \frac{1}{2} + \frac{1}{2}(1 + E[X]). \qquad (5.190)$$

从式(5.190)中我们得到$E[X] = 2$, 即扔硬币直到得到正面的期望次数是2.

当硬币正面和背面的概率不等, 得到正面概率为p时, 类似的方法仍然适用. 式(5.190)变为

$$E[X] = p + (1 - p)(1 + E[X]).$$

我们得到 $E[X] = \frac{1}{p}$. □

题目6. 扔一枚公平的硬币直到连续得到两个正面, 需要的次数的期望是多少?假如这枚硬币得到正面的概率是25%呢?

参考答案: 我们直接计算硬币不公平的情况, 假设这枚硬币得到正面的概率为p.

前两次扔硬币的可能结果为:

- 第一次扔出背面, 概率为$1 - p$, 这种情况下将使期望次数增加1.
- 前两次都是正面, 概率为p^2, 则扔硬币次数为2.
- 第一次扔出正面, 第二次扔出背面, 概率为$p(1 - p)$, 这种情况将使扔硬币的期望次数增加2.

如果$E[X]$表示扔出连续两次正面的期望次数，我们得到

$$E[X] = (1-p)(1+E[X]) + 2p^2 + p(1-p)(2+E[X]).$$

解出$E[X]$，我们得到

$$E[X] = \frac{1+p}{p^2}. \tag{5.191}$$

对于公平硬币，即$p = \frac{1}{2}$，由(5.191)式我们得到$E[X] = 6$，即得到连续两个正面的期望扔硬币次数为6.

对于扔出正面概率为25%的硬币，即$p = \frac{1}{4}$，由(5.191)式得到$E[X] = 20$，即得到连续两个正面的期望次数为20. □

题目7. 扔一枚公平的硬币n次，求不会连续两次出现正面的概率?

参考答案: 由正面和背面组成的长度为n的序列总共有2^n个. 若在这2^n个序列中，有a_n个序列不存在两个连续正面. 则没有连续两个正面的概率为 $\frac{a_n}{2^n}$.

注意到 $a_1 = 2$ (扔硬币一次，无论正面还是背面都不存在两个连续的正面)以及 $a_2 = 3$ (四种情况为正正，正背，背正，背背，只有正正有两个连续的正面). 接下来我们推导a_n的递推关系:

对于任何长度为n $(n \geq 3)$的序列，不存在连续2个正面的情况只有两种: (i) 它以一个背面开始，接上一个长度为$n-1$且没有两个连续正面的序列； (ii) 或者以一正、一背开始，接上一个长度为$n-2$且没有两个连续正面的序列. 由于这两种情况是互斥的，我们得到

$$a_n = a_{n-1} + a_{n-2}, \ \forall\, n \geq 3. \tag{5.192}$$

注意到(5.192)实际上是斐波那契数列的递推式.记 $\phi_1 = \frac{1+\sqrt{5}}{2}$和$\phi_2 = \frac{1-\sqrt{5}}{2}$ 为式(5.192)的特征方程$x^2 - x - 1 = 0$的根，则

$$a_n = C_1\phi_1^n + C_2\phi_2^n, \ \forall\, n \geq 1,$$

其中常数C_1和C_2由 $a_1 = 2$和$a_2 = 3$解出.

求解线性方程组

$$\begin{cases} C_1\phi_1 + C_2\phi_2 & = & 2 \\ C_1\phi_1^2 + C_2\phi_2^2 & = & 3, \end{cases}$$

我们得到

$$\begin{aligned} C_1 &= \frac{3+\sqrt{5}}{2\sqrt{5}} = \frac{\phi_1^2}{\sqrt{5}}; \\ C_2 &= \frac{3-\sqrt{5}}{2\sqrt{5}} = -\frac{\phi_2^2}{\sqrt{5}}. \end{aligned}$$

最终得到

$$a_n = \frac{1}{\sqrt{5}}\left(\phi_1^{n+2} - \phi_2^{n+2}\right),$$

因此掷n枚公平硬币, 没有两个连续正面出现的概率是

$$\frac{1}{2^n\sqrt{5}}\left(\phi_1^{n+2} - \phi_2^{n+2}\right). \quad \square$$

题目8. 你有两只相同的法贝热彩蛋, 如果从100层的高楼上扔下来都会摔碎.你的任务是确定使得彩蛋扔下来不会摔碎的最高楼层(以下称为"耐碎楼层"). 最少需要扔多少次才能完成任务?在尝试的过程中, 两只彩蛋都允许摔碎.

参考答案: 考虑如下更一般性的题目:

记$h_e(n)$为使用e只彩蛋最多一共扔n次的情况下,可以确定耐碎楼层的最大总楼层数(译者注:即如果高楼有$h_e(n)+1$层,e只彩蛋最多一共扔n次将不足以保证能找到耐碎楼层).求$h_e(n)$.

由于扔一次彩蛋只能确定一层楼, 可知

$$h_e(1) = 1. \tag{5.193}$$

另一种情况, 如果我们只有一只蛋, 那唯一可行的策略是从第一层到最高层一层一层的尝试, 所以

$$h_1(n) = n.$$

当$e \geq 2$和$n \geq 2$时, 第一次尝试最高只能在 $h_{e-1}(n-1)+1$层. 因为如果第一次尝试中彩蛋就摔碎了, 我们就只剩下$e-1$只蛋和$n-1$次尝试, 我们接下来能确定的最高总楼层就是$h_{e-1}(n-1)$了. 如果第一次尝试中, 彩蛋并未摔碎, 我们就可以把原先的$h_{e-1}(n-1)+2$层当做新的第1层, 然后题目简化为有e个蛋和$n-1$次尝试. 因此, 我们得到等式

$$h_e(n) = 1 + h_{e-1}(n-1) + h_e(n-1).$$

不断展开递推式, 我们得到

$$\begin{aligned}
& h_e(n) \\
= \;& 1 + h_{e-1}(n-1) + h_e(n-1) \\
= \;& 2 + h_{e-1}(n-1) + h_{e-1}(n-2) + h_e(n-2) \\
= \;& \ldots \\
= \;& (n-1) + \sum_{j=1}^{n-1} h_{e-1}(j) + h_e(1) \\
= \;& n + \sum_{j=1}^{n-1} h_{e-1}(j),
\end{aligned}$$

我们使用了$h_e(1)=1$, 如(5.193)式.

对于题目中的$e=2$, 上式变为

$$\begin{aligned}
& h_2(n) \\
= \;& n + \sum_{j=1}^{n-1} h_1(j) = n + \sum_{j=1}^{n-1} j = n + \frac{(n-1)n}{2} \\
= \;& \frac{n(n+1)}{2}, \qquad (5.194)
\end{aligned}$$

其中倒数第二步使用了$\sum_{j=1}^{k} j = \frac{k(k+1)}{2}$.

由于 $h_2(13)=91 < 100 \leq 105 = h_2(14)$, 需要扔彩蛋的最少次数为14.

注意到我们的迭代式也提供了一种算法: 将第一只彩蛋分别在这些楼层扔下

$$\begin{aligned}
14 &= 1 + h_1(13), \\
27 &= 2 + h_1(13) + h_1(12), \\
39 &= 3 + \sum_{j=11}^{13} h_1(j), \\
50 &= 4 + \sum_{j=10}^{13} h_1(j), \\
60 &= 5 + \sum_{j=9}^{13} h_1(j), \\
69 &= 6 + \sum_{j=8}^{13} h_1(j), \\
77 &= 7 + \sum_{j=7}^{13} h_1(j), \\
84 &= 8 + \sum_{j=6}^{13} h_1(j), \\
90 &= 9 + \sum_{j=5}^{13} h_1(j), \\
95 &= 10 + \sum_{j=4}^{13} h_1(j), \\
99 &= 11 + \sum_{j=3}^{13} h_1(j),
\end{aligned}$$

和100; 即首先向上移动$14 = 1 + h_1(13)$层, 然后移动$13 = 1 + h_1(12)$层, 再移动$12 = 1 + h_1(11)$层, 以此类推, 直到第一只彩蛋摔碎(或者到100层都没有摔碎). 记第一只蛋摔碎的楼层为f, 以及上一次扔彩蛋的楼层f', 将第二只蛋依次从

中间的楼层扔下, 即 $f'+1, f'+2, \ldots, f-1$. □

题目9. 一只蚂蚁在一间$10\times10\times10$的房间的一个角落.它想爬到体对角的最短路径是多长?

参考答案: 为了表述方便, 我们假定蚂蚁现在在天花板的一角, 记为A; 记A的体对角为B (在地板上).从A到B的最短路径显然需要从墙上走一条直线到地板的一边, 然后再在地板上走一条直线到B. 我们想象将蚂蚁经过的那面墙放倒, 得到一个10×20的长方形, 而A和B正好在长方形相对的两个角. 从A到B的最短路径就是长方形的对角线, 其长度为$10\sqrt{5}\approx22.36$. □

题目10. 一个$10\times10\times10$的正方体由$1,000$个单位正方体组成.你可以从正方体的表面看到多少个单位正方体?

参考答案: 如果将所有最外层的单位正方体移除, 剩下的是一个 $8\times8\times8$的正方体, 包含$8^3=512$个单位正方体. 因此, 最外层有 $1000-512=488$个单位正方体. □

题目11. 福克斯·穆德被外星人关在一个圆形围栏里.在围栏外有一名外星人守卫, 它跑步的速度是穆德的四倍, 但它仅限于在围栏周围活动并无法进入围栏. 如果穆德能成功地跑到围栏边并且守卫未赶到, 他能迅速的把围栏间的缝变大并逃走, 他能否在守卫到达之前跑到这样的位置呢?

参考答案: 令R表示围栏的半径, C表示圆心.记穆德的速度为v, 外星人的速度为$4v$.将穆德和外星人的位置分别记为M和A. 穆德首先不能直接从圆心C跑到围栏边, 因为穆德需要$\frac{R}{v}$的时间跑过R的距离, 而外星人早在 $\frac{\pi R}{4v}$的时间里就能跑过整个半圆, 到对面等着穆德了.

为了优化策略, 穆德应该在一个比圆心C更靠近围栏边的位置开始跑向围栏.假设穆德成功的到达与C距离xR的点M(其中 $0<x<1$), 其中M, C和A在同一条直线上, 且C在M和A之间.记在圆上A对面的点为P, 如图5.2. 则$MC=xR$和$MP=(1-x)R$. 穆德需要$\frac{(1-x)R}{v}$时间从M到围栏边

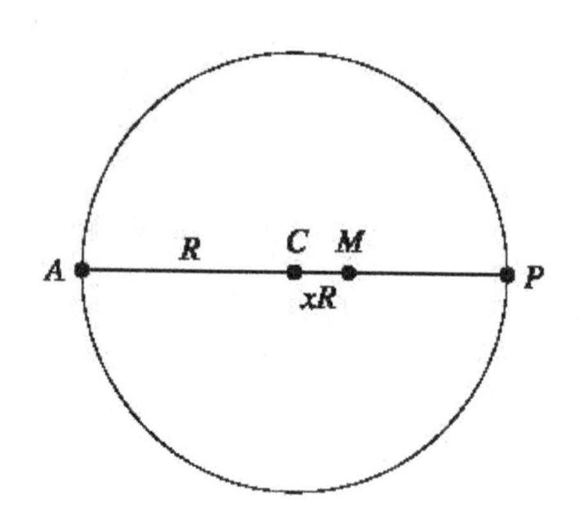

图 5.2: 穆德能在外星人到达前从M点到P点.

缘P, 而外星人需要$\frac{\pi R}{4v}$时间沿半圆从A到P. 注意到

$$\frac{(1-x)R}{v} < \frac{\pi R}{4v}, \text{ if } x > 1 - \frac{\pi}{4}.$$

因此, 如果能够使$x > 1 - \frac{\pi}{4}$, 穆德就能成功地从外星人手里逃脱.

我们现在来阐述穆德逃跑的策略.首先令$x = 1 - \frac{\pi}{4} + 0.01$.注意到由于 $0.01 < \frac{\pi-3}{4}$, $x < \frac{1}{4}$.不管外星人如何行动, 穆德先从C跑到以C为圆心, 半径为xR的圆上任意一点.然后他沿着这个圆跑, 直到他的位置M正好与A在圆心的对面.这点他是能够做到的, 因为$x < \frac{1}{4}$而外星人的速度仅仅是他的4倍 [13]. 最后, 穆德再从M点跑向P点, 并比外星人更早到达P点, 因为如上所述, $x > 1 - \frac{\pi}{4}$. □

---

[13] 由于$x < \frac{1}{4}$, 穆德的角速度$\frac{\theta v}{\pi R x}$比外星人的角速度$\frac{4\theta v}{\pi R}$大:

$$\frac{\theta v}{\pi R x} > \frac{4\theta v}{\pi R} \iff \frac{1}{4} > x.$$

题目12. 在你常去的地铁站, 你注意到两个方向的地铁到达的频率应当是一样的, 然而80%的时间里往一个方向的火车先到达, 而往另一个方向的火车只有20%的时间先到达.这是怎么回事呢?

参考答案: 一种可能的情况是80%时间先到达的那条地铁线确实更加频繁.但是, 即使两辆列车的频率相同, 其中一条地铁线也可能在80%的时间里先到达. 比如, 假定你进地铁站的时间是均匀分布, 如果地铁每十分钟到达一次, 而地铁A到站时间为1:00, 1:10, 1:20, ..., 而地铁B到站时间为1:12, 1:22, 1:32, ..., 那么地铁A就会在80%的时间里更早到达. □

题目13. 一开始我们有一只阿米巴虫.每分钟这只阿米巴虫要么死亡, 要么什么都不做, 要么分裂成两只, 要么分裂成三只. 这四种情况以等概率发.以后所有分裂得到的阿米巴虫行为也一样. 求阿米巴虫最终全部死亡的概率?

参考答案:
解答1: 记A_k为在k分钟后没有任何阿米巴虫存活的事件.令$p_k = P(A_k)$. 注意到 $p_1 = \frac{1}{4}$, 以及对任意$k \geq 1$ $A_k \subseteq A_{k+1}$.

阿米巴虫全部死亡的概率p是存在某个时间点, 在此之后没有任何阿米巴虫存活的概率.换言之,

$$p = P\left(\bigcup_{k=1}^{\infty} A_k\right).$$

注意到, 由于对于任意$k \geq 1$, $A_k \subseteq A_{k+1}$, $\bigcup_{k=1}^{n} A_k = A_n$.

则

$$\begin{aligned} p &= P\left(\bigcup_{k=1}^{\infty} A_k\right) = P\left(\lim_{n\to\infty} \bigcup_{k=1}^{n} A_k\right) \\ &= \lim_{n\to\infty} P\left(\bigcup_{k=1}^{n} A_k\right) = \lim_{n\to\infty} P\left(A_n\right) \\ &= \lim_{n\to\infty} p_n. \end{aligned} \quad (5.195)$$

由于第一分钟后有四种等概率的可能情况, 即阿米巴虫死亡、不变、分裂为两只、分裂为三只, 可知对于任意$n \geq 2$,

$$p_n = \frac{1}{4} + \frac{1}{4}p_{n-1} + \frac{1}{4}p_{n-1}^2 + \frac{1}{4}p_{n-1}^3, \tag{5.196}$$

这个序列$(p_n)_{n\geq 0}$是单调递增的: 由于$A_n \subseteq A_{n+1}$, 因此

$$p_n = P(A_n) \leq P(A_{n+1}) = p_{n+1}, \ \forall\, n \geq 1.$$

另外, 根据归纳法我们得到序列$(p_n)_{n\geq 0}$的上界是$\sqrt{2}-1$, 这是由于$p_1 = \frac{1}{4} < \sqrt{2}-1$, 以及, 如果我们假定对于某个$n \geq 2$, $p_{n-1} < \sqrt{2}-1$, 则由式(5.196), 我们得到

$$\begin{aligned} p_n &< \frac{1}{4} + \frac{1}{4}(\sqrt{2}-1) + \frac{1}{4}(\sqrt{2}-1)^2 + \frac{1}{4}(\sqrt{2}-1)^3 \\ &= \sqrt{2}-1. \end{aligned}$$

所以, 序列$(p_n)_{n\geq 0}$递增且存在上界, 因此是收敛的.

再由式(5.195)知 $\lim_{n\to\infty} p_n = p$, 其中 p是阿米巴虫最终全部死亡的概率. 由对于任意$n \geq 1$, $p_n < \sqrt{2}-1$, 可知$p \leq \sqrt{2}-1$. 再由式(5.196), 我们得到

$$p = \frac{1}{4} + \frac{1}{4}p + \frac{1}{4}p^2 + \frac{1}{4}p^3,$$

可以化为

$$0 = p^3 + p^2 - 3p + 1 = (p-1)(p^2 + 2p - 1),$$

方程的解为1, $\sqrt{2}-1$, $-\sqrt{2}-1$. 由于 $0 < p \leq \sqrt{2}-1$, 我们得到 $p = \sqrt{2}-1$.

所以最终得到阿米巴虫全部死亡的概率为$\sqrt{2}-1$.

解答2: 令p为一只阿米巴虫后代全部死亡的概率, 由于每只阿米巴虫都是相互独立的, n只阿米巴虫的后代全部死亡的概率为p^n.

另外, 当我们考虑了所有可能性之后, 一只阿米巴虫的后代全部死亡的概率是独立于时间的.

在一开始时的阿米巴虫后代全部死亡的概率，根据定义，正是p. 在一分钟内，这只阿米巴虫变成0, 1, 2或3只，概率均为$\frac{1}{4}$. 则这只阿米巴虫与其可能的后代最终全部死亡的概率为

$$\frac{1}{4}\left(p^0+p^1+p^2+p^3\right) = \frac{1}{4}\left(1+p+p^2+p^3\right).$$

两种方法算出的概率必然相等，所以

$$p = \frac{1}{4}\left(1+p+p^2+p^3\right), \tag{5.197}$$

变换为

$$p^3+p^2-3p+1 = (p-1)(p^2+2p-1) = 0. \tag{5.198}$$

式(5.198)的解为$p=1$, $p=-\sqrt{2}-1$, 和$p=\sqrt{2}-1$. 唯一在区间$(0,1)$中的解是$p=\sqrt{2}-1$, 这就是阿米巴虫最终全部死亡的概率.

一个小小的疑问在于，为什么$p = 1$不是这个题目的解?这是因为式(5.197)的右边是描述阿米巴虫繁殖的分支过程的生成函数$h(p)$. 一个广为人知的关于分支过程的定理告诉我们，如果一只阿米巴虫分裂数目的均值大于1，则方程$p=h(p)$最小的正根是阿米巴虫全部死亡的概率. 在我们的题目中，阿米巴虫分裂数目的均值是$\frac{1}{4}(0+1+2+3) = \frac{3}{2} > 1$, 因此这个定理是适用的. □

题目14. 给定一个集合X, 其中有n个元素，随机地选择两个子集A和B. 求A是B的一个子集的概率?

参考答案: 当随机选择X的两个子集A和B时，每个元素以相等概率属于以下四个集合:

$$A\setminus B,\ \ B\setminus A,\ \ A\cap B,\ \ \text{and}\ \ X\setminus(A\cup B).$$

为了使A是B的一个子集，$A\setminus B$必须是空集；换言之，X中没有任何一个元素在$A\setminus B$中.而对每一个元素，这件事的概率是$\frac{3}{4}$.

我们因此求出A是B的一个子集的概率为

$$\left(\frac{3}{4}\right)^n. \quad \square$$

题目15. 爱丽丝在两张纸上各写了一个0和1之间不同的实数, 鲍勃随机选择一张纸, 看上面的数, 然后要说出他看的纸上的数是两个数中较大还是较小的一个.

有没有任何办法能使鲍勃的正确率超过一半呢?

参考答案: 记爱丽丝在两张纸上写的数为a_1和a_2, $0 < a_1 < a_2 < 1$. 假设鲍勃看到的数是A, 他的任务是 (以一半以上的正确率)猜测$A = a_1$还是$A = a_2$.

鲍勃的策略如下: 当看到A时, 鲍勃以均匀分布随机的从区间$(0,1)$选择一个数B, 如果B比A小, 则猜$A = a_2$, 反之猜$A = a_1$.

用E表示鲍勃使用这个策略猜测正确的事件, 则

$$\begin{aligned}
P(E) &= P(E|A = a_1) \cdot P(A = a_1) \\
&\quad + P(E|A = a_2) \cdot P(A = a_2) \\
&= P(B > a_1) \cdot P(A = a_1) \\
&\quad + P(B < a_2) \cdot P(A = a_2) \\
&= (1 - a_1) \cdot 0.5 + a_2 \cdot 0.5 \\
&= 0.5 + 0.5(a_2 - a_1) \\
&> 0.5.
\end{aligned}$$

因此的概率猜测正确的概率大于 $\frac{1}{2}$. $\square$

题目16. 125^{100}一共有几位?不允许使用 $\log_{10} 2$ 或者 $\log_{10} 5$的值.

参考答案: 注意到

$$125^{100} = \left(\frac{1000}{8}\right)^{100} = \frac{1000^{100}}{2^{300}}. \tag{5.199}$$

又因为 $2^{10} = 1024$则

$$125^{100} = \frac{1000^{100}}{1024^{30}} = \frac{1000^{70}}{1.024^{30}}. \quad (5.200)$$

我们首先证明

$$1 < 1.024^{30} < 10. \quad (5.201)$$

使用二项式定理展开则有

$$(1+0.024)^{30} = \sum_{j=0}^{30} \binom{30}{j} 0.024^j. \quad (5.202)$$

我们发现,展开(5.202)后相邻两项的比率小于0.72. 证明如下:

$$\begin{aligned} \frac{\binom{30}{j+1}0.024^{j+1}}{\binom{30}{j}0.024^j} &= \frac{\binom{30}{j+1}}{\binom{30}{j}} \cdot 0.024 \\ &= \frac{30!}{(j+1)!\,(30-j-1)!} \cdot \frac{j!\,(30-j)!}{30!} \cdot 0.024 \\ &= \frac{30-j}{j+1} \cdot 0.024 \\ &< 30 \cdot 0.024 \\ &= 0.72. \end{aligned}$$

那么,

$$\binom{30}{j} 0.024^j \le 0.72^j \ \forall\, 0 \le j \le 30, \quad (5.203)$$

通过 (5.202) 和 (5.203), 我们得到

$$\begin{aligned} (1+0.024)^{30} &< \sum_{j=0}^{30} 0.72^j < \sum_{j=0}^{\infty} 0.72^j \\ &= \frac{1}{1-0.72} \\ &< 10; \end{aligned}$$

上面倒数第二步我们用到了几何序列的极限公式.

$$\sum_{j=1}^{\infty} z^j = \frac{1}{1-z} \text{ for } z = 0.72.$$

从而不等式 (5.201) 得证.
又通过 (5.200) 和 (5.201), 我们得到

$$10^{209} < 125^{100} = \frac{10^{210}}{1.024^{30}} < 10^{210}, \tag{5.204}$$

从而我们知道125^{100} 有 210 位. □

题目17. 给集合$\{1, 2, 3, \ldots, 2013\}$的每个子集规定一个权重. 计算权重的方式是把子集中的数从小到大排列,然后在这串数中交替添加"减"和"加"[14]. 求
$\{1, 2, 3, \ldots, 2013\}$ 的所有子集的权重之和.

参考答案: 记 $w(H)$ 为子集合H的权重. 不包含元素1的任意子集S,可以与包含1的子集$\{1\} \cup S$配对.因为
$\{1, 2, 3, \ldots, 2013\}$的子集个数为$2^{2013}$,所以一共有$2^{2012}$对.并且 $w(S) + w(\{1\} \cup S) = 1$;例如

$$\begin{aligned} & w(\{2,5,8\}) + w(\{1,2,5,8\}) \\ = & (2 - 5 + 8) + (1 - 2 + 5 - 8) \\ = & 1. \end{aligned}$$

从而我们知道$\{1, 2, 3, \ldots, 2013\}$所有子集的权重之和为$2^{2012}$.
□

题目18. A和B轮流从下列9个数中不放回地选一个数: 1/16, 1/8, 1/4, 1/2, 1, 2, 4, 8, 16. 谁先能在抽到的数中选三个数乘起来为1就获胜.如果A先开始选,她的策略是什么?她一定能赢吗?

参考答案: 注意到这9个数都可以写成2的整数次方: 2^{-4}, 2^{-3}, 2^{-2}, 2^{-1}, 2^0, 2^1, 2^2, 2^3, 和 2^4. 其次如果我们把这些

[14]例如, 子集 $\{3\}$ 的权重为 3. $\{2, 5, 8\}$ 子集权重为$2 - 5 + 8 = 5$.

数排列成一个 3×3矩阵,列向量分为为(从左到右):第一行是$2^3, 2^{-4}, 2^1$; 第二行是 $2^{-2}, 2^0, 2^2$;第三行是 $2^{-1}, 2^4, 2^{-3}$.

这样组成的矩阵的每行、每列和对角线之积均为1.那么原题目就转变成了井字过三关(Tic-Tac-Toe)题目. 我们知道这种游戏,先开始的人总可以不输.请向您身边的高玩请教不败之道吧! □

题目19. J夫妇邀请了4对夫妇来家里做客.晚宴结束时,J夫人询问在场的每一个人(包括J先生)跟几个人握过手. 她得到了9个不同的答案.我们假设夫妇之间不握手,并且跟同一个人不会握两次手.请问J先生一共跟几个人握过手?

参考答案: 记P_i 为跟i个人握过手的人, $i=0,1,\ldots,8$.这里我们把J夫人先排除在外进行分析. 那么P_8跟8个人握过手,而在场一共10个人,那么他/她唯一没有握过手的人就是他/她的配偶. 又P_0没有跟任何人握过手,而P_8唯一没有握过手的人是自己的配偶,所以 P_0和 P_8是夫妇.

P_7 跟两个人没有握过手,其中一个是自己的配偶,另外一个是 P_0.同理 P_1仅仅跟一个人握了手跟她/他握过手的人一定是 P_8,那么 P_1没跟 P_7握过手,所以 P_1和 P_7是一对. 以此类推,我们发现: P_6 和 P_2 是一对,P_5 和 P_3 也是两口子,这样P_4就一定是J先生.所以J先生一共跟4个人握过手.

□

题目20. 纽约洋基队和旧金山巨人队在棒球世界锦标赛的决赛对决(BO7). 你想赌$100洋基队最后获得冠军, 但只能下注每一场比赛的输赢.请问第一场比赛应当下多少注?

参考答案: 记 P为 5×5 矩阵,其中$P(i,j)$ 表示当洋基队胜利i场输了 j场比赛时我们的盈利. $(0\le i,j\le 4)$. 很明显哪只队伍先赢四场就夺得冠军,即 $P(4,j)=100$ $(0\le j\le 3)$,并且 $P(i,4)=-100$ $(0\le i\le 3)$. 而且$P(4,4)$ 为无效数字因为最多只打7场比赛.

记 B 为4×4 矩阵,每一个坐标 $B(i,j)$表示洋基队胜利i场输了 j场比赛的情况下,我们应当在下一场赌多少钱.很显然因为我们最终希望洋基队获得总冠军的时候一共赢$100,则第七场的时候我们一定要下注$B(3,3)=100$.

当洋基队胜利i场输了 j场比赛时如果我们下场比赛赌上$B(i,j)$,当洋基队赢了下一场我们的收益是 $P(i,j)+B(i,j)$ 但是如果洋基队在下一场输了,则我们的收益为 $P(i,j)-B(i,j)$. 所以有,

$$\begin{aligned} P(i+1,j) &= P(i,j)+B(i,j); & (5.205) \\ P(i,j+1) &= P(i,j)-B(i,j). & (5.206) \end{aligned}$$

等式变换之后有:

$$\begin{aligned} P(i,j) &= \frac{1}{2}(P(i+1,j)+P(i,j+1)) & (5.207) \\ B(i,j) &= \frac{1}{2}(P(i+1,j)-P(i,j+1)) & (5.208) \end{aligned}$$

这样我们就得到了一个递推公式,根据上面我们分析的边界条件我们可以得到

$$P(3,3) = \frac{1}{2}(P(4,3)+P(3,4)) = \frac{1}{2}(100-100)=0.$$

一旦我们完成了矩阵 P 通过 (5.208) 就可以计算出矩阵B的数值. 例如:

$$B(2,1) = \frac{1}{2}(P(3,1)-P(2,2)) = \frac{1}{2}(75-0)=37.5.$$

换句话说如果洋基队已经赢了 2场比赛,输了1比赛,我们就需要在下场比赛中赌 \$37.5 . 最终我们得到结果:

$$P = \begin{pmatrix} 0 & -31.25 & -62.5 & -87.5 & -100 \\ 31.25 & 0 & -37.5 & -75 & -100 \\ 62.5 & 37.5 & 0 & -50 & -100 \\ 87.5 & 75 & 50 & 0 & -100 \\ 100 & 100 & 100 & 100 & 0 \end{pmatrix};$$

$$B = \begin{pmatrix} 31.25 & 31.25 & 25 & 12.5 \\ 31.25 & 37.5 & 37.5 & 25 \\ 25 & 37.5 & 50 & 50 \\ 12.5 & 25 & 50 & 100 \end{pmatrix}.$$

因此查表我们得到第一场比赛我们应当赌 $B(0,0) = 31.25$ 元. 注意到第一场过后不管洋基队的输赢结果我们对于第二场下注的量相同. □

题目21. 桌上摆放着两个红球,两个绿球和两个黄球.每种颜色有一个重球,一个轻球.不同颜色的重球质量相同,不同颜色的轻球质量也相同.用一个没有刻度的天平, 最少需要几次称量才能找出三个重球?

参考答案: 一次称重显然无法分辨,那么我们下面就证明如何仅仅使用两次称重就可以分辨出哪三个球是较重的. 如果我们分别把这些球编上号,红球分别为 R_1 和 R_2,绿球分别为 G_1 和 G_2,黄球为 Y_1 和 Y_2. 第一次我们称量 $\{R_1, G_1\}$ 和 $\{R_2, Y_1\}$. 如果天平保持平衡,那么 G_1 或者 Y_1 其中有且只有一个是重球. 在这种情况下我们第二次称量 $\{G_1\}$ 和$\{Y_1\}$. 如果 G_1更重那么重球分别为$\{R_2, G_1, Y_2\}$. 但如果 Y_1 更重,则重球为 $\{R_1, G_2, Y_1\}$.

如果第一次称量时$\{R_1, G_1\}$ 更重,那么要不然G_1 是重球要不然 Y_1 是轻球.那么我们第二次称量 $\{G_1, Y_1\}$ 和 $\{G_2, Y_2\}$.如果天平平衡,那么G_1 是重球从而重球的编号为 $\{R_1, G_1, Y_2\}$. 如果 $\{G_1, Y_1\}$ 更重,那么 G_1 和 Y_1 都是重球.由此可知重球的编号分别为 $\{R_1, G_1, Y_1\}$. 如果 $\{G_2, Y_2\}$ 更重,那么 G_2 和 Y_2 都是重球.由此可知重球的编号则为 $\{R_1, G_2, Y_2\}$.

回到第一次称重时,如果 $\{R_2, Y_1\}$ 更重,则或者 Y_1 是重球或者G_1 是轻球.那么我们第二次称重则选择 $\{G_1, Y_1\}$ 和 $\{G_2, Y_2\}$比较.如果天平平衡,那么 Y_1更重.因此重球的编号为$\{R_2, G_2, Y_1\}$. 如果$\{G_1, Y_1\}$更重,那么 G_1 和 Y_1 都是重球.则重球编号为 $\{R_2, G_1, Y_1\}$. 如果$\{G_2, Y_2\}$ 更重那么G_2 和 Y_2 都是重球,则重球编号为 $\{R_2, G_2, Y_2\}$. □

题目22. 一字排开的十个房间中有一个藏有宝箱. 幽灵每晚把宝箱移动到一个相邻的房间.如果你每天只能查看一个房间, 那么需要使用怎样的策略才能找到宝箱?

参考答案: 宝箱最多16天后就一定会被找到.我们先将十个房间编号为 1 号到 10号. 设 T_k 为第k天宝箱所在

的房间编号,R_k为第k天你检查的房间. 使用如下的策略: 当$k = 1, 2, \ldots, 8$, 令$R_k = k + 1$. 即第一天从第二个间房开始每天按照顺序向相邻房间依次寻找. 当$k = 9, 10, \ldots, 16$, 令$R_k = 18 - k$.即第9天跟第8天检查一样的房间,然后依次向回寻找.

如果T_1是偶数,那么我们将会在前8天中的某一天找到宝箱.换句话说,存在k满足$1 \leq k \leq 8$ 使得 $T_k = R_k$.下面我们给出证明:当T_1是偶数时,因为$R_1 = 2$,所以T_1和R_k是同奇偶的.又 T_k 和 R_k每天最多变化为1,所以$T_k - R_k$为偶数.同时我们知道$T_1 \neq 1$ 并且 $T_8 \neq 10$ 意味着$T_1 - R_1 \geq 0$并且$T_8 - R_8 \leq 0$.因为$T_k - R_k$最多每天可以变化2,那么必须存在一个k 满足$1 \leq k \leq 8$使得$T_k - R_k = 0$.

如果T_1是奇数,那么当满足$9 \leq k \leq 16$时一定存在k 满足$T_k = R_k$. 因为如果T_1 是奇数,那么T_9 一定也是奇数.又因为$R_9 = 9$,所以$9 \leq k \leq 16$时 T_k 和 R_k有相同的奇偶性.那么同理$R_k - T_k$ 是偶数.而且我们很容易得到的是 $T_9 \neq 10$ 并且 $T_{16} \neq 1$,那么我们可以推理出如下关系:$R_9 - T_9 \geq 0$ 并且 $R_{16} - T_{16} \leq 0$. 因为 $R_k - T_k$每天最多变化 2 ,那么一定存在某个k满足 $9 \leq k \leq 16$ 使得$R_k - T_k = 0$.

最后,这种策略可以扩展到 $n \geq 2$个任意房间数.并且宝箱一定会在$2n - 4$ 天前被找到. 而我们检查房间的顺序是:$2, 3, \ldots, n - 1, n - 1, n - 2, \ldots, 3, 2$.

□

题目23. 请问一共要经过几次比较才能找到n个互不相同的数的最大值?如果要同时找出最大值和最小值, 一共需要多少次比较呢?

参考答案: 最简单的方法是从左到右一次比较所有数字从而得到序列最大数,共需要$n-1$ 次比较.写成伪代码:设 M为当前最大值,开始时令$M = x_1$,那么在第i次比较时比较M 和 x_{i+1}的大小,如果 $M > x_{i+1}$ 则M不变否则$M = x_{i+1}$.重复上面的步骤当$i = 1, \ldots, n - 1$.

如果需要同时找到最大和最小值,那么最简单的方法是使用 $2n-3$个比较.具体方法是使用$n-1$次比较找出最大值,然后在剩下的$n-1$个数中使用相同的方法找出最小值.

然而,对于同时找最大最小值我们有更加快捷和聪明的方法.当n是偶数的时候,比较所有$\frac{n}{2}$ 对相邻的数字,即比较x_{2i-1}和 x_{2i}的大小其中$i=1:\frac{n}{2}$. 然后把小的数放到集合 S中,把大的数放到L中.这一共需要$\frac{n}{2}$ 次比较.并且S 和 L分别有$\frac{n}{2}$个元素.那么同时找到最大最小值的题目就变化成了在S中找出最小值,在L 找出最大值.分别需要使用 $\frac{n}{2}-1$次比较.那么总共需要

$$\frac{n}{2}+\left(\frac{n}{2}-1\right)+\left(\frac{n}{2}-1\right) = \frac{3n}{2}-2. \qquad (5.209)$$

如果 n 是奇数,那么随机把一个数拿出来,按照上述同样的方法分成$\frac{n-1}{2}$对比较之后,得到S 和L两个集合就需要$\frac{n-1}{2}$次比较. 那么找出 S 的最小值和L的最大值分别需要$\frac{n-1}{2}-1$次比较. 最后再把拿出来的数分别跟最大值最小值比较比较(一共需要两次比较).那么一共需要

$$\frac{n-1}{2}+\frac{n-1}{2}-1+\frac{n-1}{2}-1+2 = \frac{3n+1}{2}-2. \quad (5.210)$$

注意到结果 (5.209) 和 (5.210) 可以写成如下形式

$$\left\lceil\frac{3n}{2}\right\rceil-2 \quad ,$$

其中 $\lceil x\rceil$ 表示向上取整 x, 即大于或等于 x的最小整数. □

题目24. 一只蚂蚁在立方体的一个顶点上. 每一步它有等概率移动到相邻的三个顶点之一. 请问它移动到对角顶点的期望步数是多少?

参考答案: 把这个立方体的每个顶点依次用相对于对角顶点的距离进行标号,所在顶点标为3, 相邻顶点标为2,再远一些的点标为1,对角顶点标为 0.

记E_i, $i = 0,1,2,3$为站在标号为i的顶点时到达标号为0的顶点的期望步数.

我们已知 $E_0 = 0$,需要计算 E_3.第一步之后,不管蚂蚁选择哪个方向,都会到达标号为2的顶点,所以

$$E_3 = 1 + E_2. \tag{5.211}$$

从标号为 2的顶点出发,有3条棱可以选择,其中两条棱的终点是标号为1的顶点,另外一条则是回到标号为3的顶点.由于蚂蚁等概率选择每条路径,所以它有$\frac{2}{3}$的概率移动到标号为1的顶点,有$\frac{1}{3}$ 的概率移动到标号为3的顶点.那么有如下关系:

$$E_2 = 1 + \frac{2}{3}E_1 + \frac{1}{3}E_3. \tag{5.212}$$

同理,当蚂蚁幸运地移动到标号为1的顶点时有$\frac{2}{3}$的概率移动到标号为2的顶点, 有 $\frac{1}{3}$的概率移动到标号为 0的顶点.那么我们有如下关系

$$E_1 = 1 + \frac{2}{3}E_2 + \frac{1}{3}E_0 = 1 + \frac{2}{3}E_2, \tag{5.213}$$

又已知 $E_0 = 0$.

联立方程组 (5.211–5.213), 解得 $E_1 = 7$, $E_2 = 9$,并且 $E_3 = 10$. 所以需要移动到对角顶点的平均期望步数为10. □

题目25. 从0到1的均匀分布中随机生成数字,若生成的数字比上一轮的数字小就继续生成,否则停止.

(i)问平均一共会生成多少个数字(生成数字总数的期望)?

(ii)问生成序列中最小数字的期望.

参考答案: 我们列出了(i)的三种解法,第三种解法将会用于解答(ii).

(i) *解法 1:* 记$E(x)$为上一个生成的数字为x的情况下,平均还能产生多少个数字.那么显而易见$E(0) = 1$,因为 $x = 0$已

经为最小的数,下次产生任何数字都不会比 0小.所以下一个数字将为最后一个数字.

假设上一个产生的数字是 x并且记 y为下一个产生的数字,我们试图计算给定y时的条件期望$E(x)$. 如果y大于x那么游戏结束,发生此情况的概率为$1-x$.当y小于x时,我们有$E(x)=1+E(y)$,概率为x.因为已知 y的概率密度函数为 $f(y)=1$, 所以整理上述逻辑我们得到如下等式:

$$\begin{aligned} E(x) &= 1\cdot(1-x)+\int_0^x (1+E(y))f(y)\,dy \\ &= (1-x)+\int_0^x (1+E(y))\,dy \\ &= 1+\int_0^x E(y)\,dy. \end{aligned}$$

整理后有:

$$E(x) = 1+\int_0^x E(y)\,dy$$

等式两边对x 求导有

$$E'(x) = E(x).$$

此常微分方程的解为$E(x)=Ce^x$, 其中 C 为常数. 又已知边界条件 $E(0)=1$, 所以 $C=1$. 最后我们得到解为:$E(x)=e^x$.

需要说明的是第一个数一定小于1,所以本题的答案是$E(1)$,代入方程得到$E(1)=e$.

(i) *解法 2:* 记N 为平均可以生成数字的个数,x_i 为所生成的第 i个数,$i\geq 1$. 同时记p_i 为$x_i<x_{i-1}<\ldots<x_1$的概率. 因为已经生成的i个数字有$i!$种不同的排列,代表这i个数字所有可能生成的顺序.但是只有其中一种顺序满足本题目的要求.所以我们能够观测到这i个数字的概率为$p_i=\frac{1}{i!}$.

已知我们在停止之前至少可以生成两个数字,那么在$i\geq 2$的情况下一共正好生成i个数字的概率等价于$x_{i-1}<\ldots<x_2<x_1$ 并且$x_i>x_{i-1}$的概率.也就是$x_{i-1}<\ldots<x_2<$

x_1的概率减去$x_i < x_{i-1} < \ldots < x_2 < x_1$的概率,即 $p_{i-1} - p_i$.

根据上述逻辑,我们有如下结果:

$$\begin{aligned} N &= \sum_{i=2}^{\infty} i(p_{i-1} - p_i) \\ &= \sum_{i=2}^{\infty} i\left(\frac{1}{(i-1)!} - \frac{1}{i!}\right) \\ &= \sum_{i=2}^{\infty} i \cdot \frac{i-1}{i!} \\ &= \sum_{i=2}^{\infty} \frac{1}{(i-2)!} \\ &= \sum_{j=0}^{\infty} \frac{1}{j!} \\ &= e, \end{aligned} \tag{5.214}$$

其中 (5.214) 可由e^x 在0点处的泰勒展开式得到.即在下式中令 $x = 0$

$$e^x = \sum_{j=0}^{\infty} \frac{x^j}{j!}, \ \forall x \in \mathbb{R},$$

.

(i) *解法 3:* 记 $p(x)\,dx$为递减序列中的有一个数字在x 到 $x + dx$区间内的概率, $p_i(x)\,dx$为第i个数字在x 到 $x + dx$区内的概率.我们有:

$$p(x)\,dx = \left(\sum_{i=1}^{\infty} p_i(x)\right) dx. \tag{5.215}$$

然后有:

$$p_i(x)\,dx = \frac{1}{(i-1)!}(1-x)^{i-1}\,dx, \tag{5.216}$$

因为$(1-x)^{i-1}$ 为前$i-1$个数字均小于x的概率.而$\frac{1}{(i-1)!}$为前$i-1$个数的顺序正好递减的概率. 由 (5.215) 和 (5.216)我们得到:

$$\begin{aligned} p(x)\,dx &= \sum_{i=1}^{\infty} \frac{(1-x)^{i-1}}{(i-1)!}\,dx \\ &= \sum_{j=0}^{\infty} \frac{(1-x)^{j}}{j!}\,dx \\ &= e^{1-x}\,dx; \end{aligned} \tag{5.217}$$

其中 (5.217)是将 e^t 在 0点处泰勒展开得到的.即在下式中令$t=1-x$:

$$e^t = \sum_{j=0}^{\infty} \frac{t^j}{j!}, \ \ \forall\, t \in \mathbb{R},$$

令$t=1-x$.

那么选出来的数字个数的期望为

$$\int_0^1 p(x)\,dx = \int_0^1 e^{1-x}\,dx = e-1.$$

最后,我们加上最后一个生成的不再递减的数字,得到e为平均生成的数字个数.

(ii) 记 s 为整个序列中最小的数字,即倒数第二个数字. 根据(5.217)我们知道数字在x 到 $x+dx$之中的概率为$e^{1-x}\,dx$,且因为下一个选出的数字比这个数字大的概率为$(1-x)$,所以倒数第二个数字在x 到 $x+dx$区间中的概率为 $e^{1-x}(1-x)\,dx$,这就是s 落在区间 x 和 $x+dx$之间的概率.

则最小数字的期望为

$$\begin{aligned} E(s) &= \int_0^1 xe^{1-x}(1-x)\,dx \\ &= \int_0^1 e^y y(1-y)\,dy, \end{aligned} \tag{5.218}$$

其中 (5.218) 是通过$y=1-x$的变量替换得到的.

对(5.218)使用分部积分我们得到:

$$\begin{aligned}
E(s) &= e^y(y-y^2)\Big|_0^1 - \int_0^1 e^y(1-2y)\,dy \\
&= e^y(y-y^2)\Big|_0^1 - e^y\Big|_0^1 + 2\int_0^1 ye^y\,dy \\
&= e^y(y-y^2-1)\Big|_0^1 + 2ye^y\Big|_0^1 - 2\int_0^1 e^y\,dy \\
&= e^y(3y-y^2-1)\Big|_0^1 - 2e^y\Big|_0^1 \\
&= e^y(3y-y^2-3)\Big|_0^1 \\
&= 3-e. \quad \square
\end{aligned}$$

题目26. 在一个需要筹集10万元的慈善晚会上,相互不认识(捐赠金额互相独立)的捐赠者一个接一个在晚会结束时把捐款投入捐款箱. 每个捐款者捐赠的数目服从均值为2万元的指数分布.当主办方发现捐款总额等于或超过10万时捐款活动停止. 请问捐赠者个数服从什么分布,这个分布的均值和方差分别是多少?

参考答案: 记 a_i 为捐赠者i所捐的数目,我们知道从 $1,\ldots,n$个捐款者得到的捐款总额为$s_n=\sum_{i=1}^n a_i$. 同时我们定义

$$N = \min_{n\geq 1}\{n \text{ s.t. } s_n \geq a\}$$

为满足s_n至少为a的最小的n,它是一个离散的随机变量. 同时我们记 $P(n|a)$, $n\geq 1$为 N的概率,即当需要筹集数目为a的资金时,$N=n$的概率.

$$P(1|a) = P(a_1\geq a) = e^{-\lambda a}. \tag{5.219}$$

假设第一个捐赠者的捐款为 $a_1=x<a$, N 等于n当且仅当剩下$a-x$万元正好被$n-1$个人募捐完毕.这个事件的概率为$P(n-1|a-x)$. 又因为a_1服从指数分布,所以其概率密度函数为$f_{a_1}(x)=\lambda e^{-\lambda x}$. 当$n>1$时我们有:

$$
\begin{aligned}
P(n|a) &= \int_0^a P(n-1|a-x) f_{a_1}(x)\, dx \\
&= \int_0^a \lambda e^{-\lambda x} P(n-1|a-x)\, dx. \quad (5.220)
\end{aligned}
$$

下面我们将通过数学归纳法证明:

$$
P(n|a) = \frac{(\lambda a)^{n-1}}{(n-1)!}\, e^{-\lambda a}, \ \ \forall\, a \geq 0, \qquad (5.221)
$$

当 $n = 1$时结论显然,见 (5.219). 假设 (5.221) 在 $n > 1$时成立,接下来我们证明其在$n + 1$时也成立.通过假设我们知道:对于所有$0 \leq x \leq a$有以下关系

$$
P(n|a-x) = \frac{(\lambda(a-x))^{n-1}}{(n-1)!}\, e^{-\lambda(a-x)}, \qquad (5.222)
$$

因为 (5.220)有:

$$
P(n+1|a) = \int_0^a \lambda e^{-\lambda x} P(n|a-x)\, dx. \qquad (5.223)
$$

又因为 (5.222) 和 (5.223)

$$
\begin{aligned}
& P(n+1|a) \\
= \ & \int_0^a \lambda e^{-\lambda x} \cdot \frac{(\lambda(a-x))^{n-1}}{(n-1)!}\, e^{-\lambda(a-x)}\, dx \\
= \ & \frac{\lambda^n e^{-\lambda a}}{(n-1)!} \int_0^a (a-x)^{n-1}\, dx \\
= \ & \frac{(\lambda a)^n}{n!}\, e^{-\lambda a}.
\end{aligned}
$$

我们得到 (5.221) 在$n + 1$时也成立,所以得证.

由(5.221)可知, N与$1 + M$有相同的分布,其中M为期望为λa的泊松分布. 所以

$$
E[N] = 1 + \lambda a; \quad \mathrm{Var}(N) = \lambda a.
$$

回到原题目,代入 $1/\lambda = \$2$ 万和 $a = \$10$万. 得到 $E[N] = 1 + \lambda a = 6$ 并且 $\mathrm{Var}(N) = \lambda a = 5$.从而我们得到平均捐款人数服从均值为6、方差为5的泊松分布. □

题目 27. 一个物体从1开始等概率地向左或向右每次移动一步.如果走到0或3则停止移动.

(i) 平均移动了多少步之后停止?

(ii) 最终停在3而不是0的概率是多少?

参考答案: 记X_n^ℓ为n时刻物体的坐标,上标表示起始点的位置,那么我们知道$X_0^\ell = \ell$,需要求出 X_n^1.记T_ℓ为从 $\ell \in \{0, 1, 2, 3\}$开始第一次到达0或者3就停止的移动步数.并且记 $t_\ell = E[T_\ell]$为T_ℓ的期望.

(i)我们需要找到 t_1.下面我们推导出t_ℓ的递归关系: 当 $\ell = 1, 2,$

$$\begin{aligned} t_\ell &= E[T_\ell] \\ &= E[T_\ell | X_1^\ell = \ell - 1] \cdot P(X_1^\ell = \ell - 1) \\ &\quad + E[T_\ell | X_1^\ell = \ell + 1] \cdot P(X_1^\ell = \ell + 1) \\ &= E[1 + T_{\ell-1}] \cdot P(X_1^\ell = \ell - 1) \qquad (5.224) \\ &\quad + E[1 + T_{\ell+1}] \cdot P(X_1^\ell = \ell + 1) \qquad (5.225) \\ &= (1 + t_{\ell-1}) \cdot \frac{1}{2} + (1 + t_{\ell+1}) \cdot \frac{1}{2}, \qquad (5.226) \end{aligned}$$

上述推导中(5.224)和(5.225)用到了下列结果:

$$\begin{aligned} E[T_\ell | X_1^\ell = \ell - 1] &= E[1 + T_{\ell-1}]; \qquad (5.227) \\ E[T_\ell | X_1^\ell = \ell + 1] &= E[1 + T_{\ell+1}]. \qquad (5.228) \end{aligned}$$

这里的(5.227) 和 (5.228) 分别代表了从 ℓ开始一旦第一步走到了$\ell - 1$ (或者$\ell + 1$),就等价于从$\ell - 1$ (或者 $\ell + 1$)开始的期望(平均移动步数)加上1,这就是为什么(5.227) 和 (5.228) 的右边有加1项.

又因为 $t_0 = t_3 = 0$, 在 (5.226) 令 $l = 1$ 和 $l = 2$, 我们分别得到 t_1 和 t_2:

$$
\begin{aligned}
t_1 &= \frac{1+t_0}{2} + \frac{1+t_2}{2} \\
&= \frac{1}{2} + \frac{1+t_2}{2};
\end{aligned}
$$

$$
\begin{aligned}
t_2 &= \frac{1+t_1}{2} + \frac{1+t_3}{2} \\
&= \frac{1+t_1}{2} + \frac{1}{2}.
\end{aligned}
$$

代入数字我们得到: $t_1 = 2$, $t_2 = 2$. 从而我们知道停止前所走的平均步数为 2.

(ii) 记 p_ℓ 为从ℓ开始最终停止在3而非0的概率.那么我们需要求 p_1. 记 τ_0^ℓ 为从ℓ出发第一次抵达0的步数, τ_3^ℓ 为从ℓ出发第一次抵达3的步数. 下面我们推导出了关于p_ℓ的递归方程: 当$\ell = 1, 2$,

$$
\begin{aligned}
p_\ell &= P\left(\tau_3^\ell < \tau_0^\ell\right) \\
&= P\left(\tau_3^\ell < \tau_0^\ell \middle| X_1^\ell = \ell+1\right) P(X_1^\ell = \ell+1) \\
&\quad + P\left(\tau_3^\ell < \tau_0^\ell \middle| X_1^\ell = \ell-1\right) P(X_1^\ell = \ell-1) \\
&= P\left(\tau_3^{\ell+1} < \tau_0^{\ell+1}\right) P(X_1^\ell = \ell+1) \\
&\quad + P\left(\tau_3^{\ell-1} < \tau_0^{\ell-1}\right) P(X_1^\ell = \ell-1) \\
&= \frac{1}{2} \cdot p_{\ell+1} + \frac{1}{2} \cdot p_{\ell-1}, \qquad (5.229)
\end{aligned}
$$

由于这是一个随机游走题目,并且移动到左右两边的概率相同,所以一旦移动后,之前移动的路径与结果无关,就好像游戏重新开始了一样.

因为 $p_0 = 0$ 并且 $p_3 = 1$, 我们分别令 (5.229)中的$l = 1$ 和$l = 2$从而得到p_1 分别为 p_2:

$$\begin{aligned} p_1 &= \frac{p_0}{2} + \frac{p_2}{2} = \frac{p_2}{2}; \\ p_2 &= \frac{p_1}{2} + \frac{p_3}{2} = \frac{p_1}{2} + \frac{1}{2}. \end{aligned}$$

则$p_1 = \frac{1}{3}$, $p_2 = \frac{2}{3}$. 那么我们知道最后停留在3而非0的概率为$\frac{1}{3}$

题目 28. 一根长为1的小木棍掉到地上断成两段,断点服从0到1的均匀分布.请问较短的那段的平均长度为多少?

参考答案:

解法 1: 由于断点服从$(0,1)$的均匀分布,那么它的概率密度$f_X(x)$ 为1,其中$0 \le x \le 1$. 记L 为较短部分的长度,那么$L = \min\,(x, 1-x)$.

$$\begin{aligned} E[L] &= \int_0^1 \min\,(x, 1-x) \cdot f_X(x)\,dx \\ &= \int_0^1 \min\,(x, 1-x)\,dx \\ &= \int_0^{1/2} x\,dx + \int_{1/2}^1 (1-x)\,dx \\ &= \frac{1}{8} + \frac{1}{8} \\ &= \frac{1}{4}. \end{aligned}$$

解法 2: 由于断点服从$(0,1)$的均匀分布,记L 为较短部分的长度. 记事件A为断点X处于$\left(0, \frac{1}{2}\right)$之间;那么$\overline{A}$为断点$X$位于$\left(\frac{1}{2}, 1\right)$. 显然$P(A) = P(\overline{A}) = \frac{1}{2}$.如果事件$A$发生,则较短部分的长度为 X.而当$\overline{A}$发生则较短部分长度为$1-X$. 那么

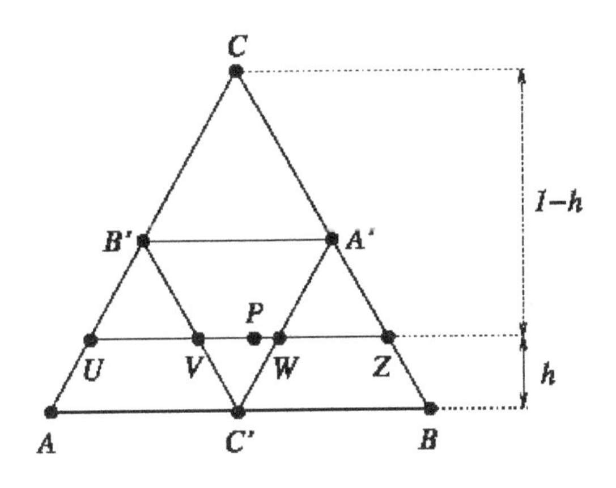

图 5.3: 点 P 是在等边三角形ABC的内接三角形 $A'B'C'$中的点.

以$\frac{VW}{A'B'} = \frac{h}{1/2}$.将这两个等式相除则有 $\frac{VW}{A'B'} = \frac{h}{1/2}$.

$$\frac{VW}{UZ} = \frac{h}{1-h}.$$

所以在$h < 1/2$的情况下,三个木段能够组成三角形的概率为

$$\int_0^{1/2} \frac{h}{1-h}\,dh = -\int_0^{1/2} \frac{1-h}{1-h}\,dh + \int_0^{1/2} \frac{1}{1-h}\,dh.$$

然后变量替换 $y = 1-h$之后则有:

$$\begin{aligned}
\int_0^{1/2} \frac{h}{1-h}\,dh &= -\frac{1}{2} + \int_{1/2}^{1} \frac{1}{y}\,dy \\
&= -\frac{1}{2} - \ln\left(\frac{1}{2}\right) \\
&= \ln 2 - \frac{1}{2}.
\end{aligned}$$

which can be written as

$$\frac{\ln(\pi)}{\pi} < \frac{\ln(e)}{e}.$$

Let $f : (0, \infty) \to \mathbb{R}$ given by $f(x) = \frac{\ln(x)}{x}$. Then,

$$f'(x) = \frac{1 - \ln(x)}{x^2}.$$

Note that $f'(x) = 0$ has one solution, $x = e$. Also, $f'(x) > 0$ for $0 < x < e$, and $f'(x) < 0$ for $x > e$, and therefore $f(x)$ is increasing on the interval $(0, e)$ and is decreasing on the interval (e, ∞).

Thus, the function $f(x) = \frac{\ln(x)}{x}$ has a global maximum point at $x = e$, i.e., $f(x) < f(e) = \frac{1}{e}$ for all $x > 0$ with $x \neq e$, and therefore

$$f(\pi) = \frac{\ln(\pi)}{\pi} < \frac{\ln(e)}{e} = \frac{1}{e},$$

which is equivalent to $\pi^e < e^\pi$; cf. (6.1). □

Question 3. Show that

$$\frac{e^x + e^y}{2} \geq e^{\frac{x+y}{2}}, \ \forall\, x, y \in \mathbb{R}. \tag{6.2}$$

Answer: Let $e^x = a$ and $e^y = b$. Note that $a, b > 0$, and that

$$e^{\frac{x+y}{2}} = \sqrt{e^{x+y}} = \sqrt{e^x \cdot e^y} = \sqrt{ab}.$$

Then, (6.2) can be written as

$$\begin{aligned} \frac{a+b}{2} \geq \sqrt{ab} \iff & a + b - 2\sqrt{ab} \geq 0 \\ \iff & \left(\sqrt{a} - \sqrt{b}\right)^2 \geq 0, \end{aligned}$$

which is what we wanted to show. □

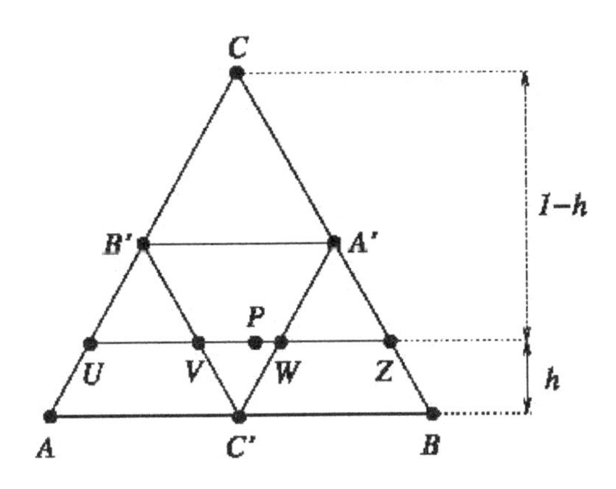

图 5.3: 点 P 是在等边三角形ABC的内接三角形 $A'B'C'$中的点.

以$\frac{VW}{A'B'} = \frac{h}{1/2}$.将这两个等式相除则有 $\frac{VW}{A'B'} = \frac{h}{1/2}$.

$$\frac{VW}{UZ} = \frac{h}{1-h}.$$

所以在$h < 1/2$的情况下,三个木段能够组成三角形的概率为

$$\int_0^{1/2} \frac{h}{1-h}\,dh = -\int_0^{1/2} \frac{1-h}{1-h}\,dh + \int_0^{1/2} \frac{1}{1-h}\,dh.$$

然后变量替换 $y = 1-h$之后则有:

$$\begin{aligned}\int_0^{1/2} \frac{h}{1-h}\,dh &= -\frac{1}{2} + \int_{1/2}^{1} \frac{1}{y}\,dy \\ &= -\frac{1}{2} - \ln\left(\frac{1}{2}\right) \\ &= \ln 2 - \frac{1}{2}.\end{aligned}$$

上面的结果是在 $h < 1/2$的情况下,而当 $h > 1/2$时候结果也是相同的,所以最后的答案是

$$\frac{\ln 2 - 1/2}{1/2} = 2\ln 2 - 1. \quad \square$$

题目 30. 为什么井盖是圆的?

参考答案: 如果要设计一个在任意方向上的宽度都不小于某个固定值的形状, 圆形是最节省材料的. 更进一步, 圆形井盖永远不会掉落到井底. 然而如果井盖是正方形的, 因为对角线长度是边长的$\sqrt{2}$倍, 所以可能掉入井底.

□

题目 31. 12点之后分针和时针在什么时间第一次重合?

参考答案: 已知分针每小时转360度, 时针每小时转30度. 12点之后, 要使两者再次相遇, 分针要比时针多走一圈, 也就是多走360度. 设下次相遇的时间为t, 则有

$$360 \cdot t = 30 \cdot t + 360.$$

解得 $t = \frac{12}{11}$. 所以大约1小时5分27秒后分针和时针再次重合. □

题目 32. 一个房间里有三盏灯, 另一个房间里有这三盏灯的开关. 能否只观察一次就找到每盏灯对应的开关?

参考答案: 在有三个开关的房间里打开两个开关, 几分钟后将其中的一个开关关闭. 接着来到有三盏灯的房间, 亮着的灯对应仍然开着的开关, 暗着并且发烫的灯对应打开几分钟后被关闭的开关, 剩下一盏暗着的灯对应从未被打开过的开关. □

Chapter 6

Solutions

6.1 Mathematics, calculus, differential equations.

Question 1. What is the value of i^i, where $i = \sqrt{-1}$?
Answer: Recall that $e^{i\theta} = \cos\theta + i\sin\theta$. Then,

$$i \;=\; \cos\frac{\pi}{2} + i\sin\frac{\pi}{2} \;=\; e^{i\frac{\pi}{2}},$$

and therefore

$$i^i \;=\; \left(e^{i\frac{\pi}{2}}\right)^i \;=\; e^{(i\frac{\pi}{2})i} \;=\; e^{i^2\frac{\pi}{2}} \;=\; e^{-\frac{\pi}{2}},$$

since $i^2 = -1$. $\square$

Question 2. Which number is larger, π^e or e^π?
Answer: We will show that $\pi^e < e^\pi$. By taking the natural logarithm, we find that

$$\begin{aligned} \pi^e < e^\pi \quad &\Longleftrightarrow \quad \ln(\pi^e) < \ln(e^\pi) \Longleftrightarrow e\ln(\pi) < \pi \\ &\Longleftrightarrow \quad \frac{\ln(\pi)}{\pi} < \frac{1}{e}, \end{aligned} \tag{6.1}$$

which can be written as

$$\frac{\ln(\pi)}{\pi} < \frac{\ln(e)}{e}.$$

Let $f : (0, \infty) \to \mathbb{R}$ given by $f(x) = \frac{\ln(x)}{x}$. Then,

$$f'(x) = \frac{1 - \ln(x)}{x^2}.$$

Note that $f'(x) = 0$ has one solution, $x = e$. Also, $f'(x) > 0$ for $0 < x < e$, and $f'(x) < 0$ for $x > e$, and therefore $f(x)$ is increasing on the interval $(0, e)$ and is decreasing on the interval (e, ∞).

Thus, the function $f(x) = \frac{\ln(x)}{x}$ has a global maximum point at $x = e$, i.e., $f(x) < f(e) = \frac{1}{e}$ for all $x > 0$ with $x \neq e$, and therefore

$$f(\pi) = \frac{\ln(\pi)}{\pi} < \frac{\ln(e)}{e} = \frac{1}{e},$$

which is equivalent to $\pi^e < e^\pi$; cf. (6.1). □

Question 3. Show that

$$\frac{e^x + e^y}{2} \geq e^{\frac{x+y}{2}}, \ \forall\, x, y \in \mathbb{R}. \tag{6.2}$$

Answer: Let $e^x = a$ and $e^y = b$. Note that $a, b > 0$, and that

$$e^{\frac{x+y}{2}} = \sqrt{e^{x+y}} = \sqrt{e^x \cdot e^y} = \sqrt{ab}.$$

Then, (6.2) can be written as

$$\begin{aligned} \frac{a+b}{2} \geq \sqrt{ab} \quad &\Longleftrightarrow \quad a + b - 2\sqrt{ab} \geq 0 \\ &\Longleftrightarrow \quad \left(\sqrt{a} - \sqrt{b}\right)^2 \geq 0, \end{aligned}$$

which is what we wanted to show. □

Question 4. Solve $x^6 = 64$.

Answer: Recall that the six unit roots of $z^6 = 1$ are

$$\begin{aligned} z_k &= \exp\left(\frac{2k\pi i}{6}\right) = \exp\left(\frac{k\pi i}{3}\right) \\ &= \cos\left(\frac{k\pi}{3}\right) + i\sin\left(\frac{k\pi}{3}\right), \end{aligned} \tag{6.3}$$

for $k = 0 : 5$. Since $\sqrt[6]{64} = 2$, we obtain from (6.3) that the solutions of $x^6 = 64$ are

$$x_k = 2\cos\left(\frac{k\pi}{3}\right) + 2i\sin\left(\frac{k\pi}{3}\right), \quad \forall\, k = 0 : 5. \quad \square$$

Question 5. What is the derivative of x^x?

Answer: Note that

$$x^x = e^{\ln(x^x)} = e^{x\ln(x)}. \tag{6.4}$$

Using Chain Rule and (6.4), we find that

$$\begin{aligned} (x^x)' &= \left(e^{x\ln(x)}\right)' = e^{x\ln(x)}\,(x\ln(x))' \\ &= x^x(\ln(x) + 1). \quad \square \end{aligned}$$

Question 6. Calculate

$$\sqrt{2 + \sqrt{2 + \sqrt{2 + \ldots}}}. \tag{6.5}$$

Answer: Assume that the limit from (6.5) exists, and denote that limit by l. Then, $l = \sqrt{2 + l}$, which can be written as

$$l^2 - l - 2 = (l-2)(l+1) = 0.$$

Since $l > 0$, we obtain that $l = 2$, i.e.,

$$\sqrt{2+\sqrt{2+\sqrt{2+\ldots}}} = 2. \tag{6.6}$$

Thus, proving (6.6) is equivalent to showing that the sequence $(x_n)_{n\geq 0}$ given by $x_0 = \sqrt{2}$ and

$$x_{n+1} = \sqrt{2+x_n}, \ \forall\, n \geq 0,$$

is convergent.

We can see by induction that the sequence $(x_n)_{n\geq 0}$ is bounded from above by 2, since $x_0 = \sqrt{2} < 2$, and, if we assume that $x_n < 2$, then

$$x_{n+1} = \sqrt{2+x_n} < \sqrt{4} = 2.$$

Moreover, the sequence $(x_n)_{n\geq 0}$ is increasing, since

$$\begin{aligned} x_n < x_{n+1} \quad &\Longleftrightarrow \quad x_n < \sqrt{2+x_n} \\ &\Longleftrightarrow \quad x_n^2 - x_n - 2 < 0 \\ &\Longleftrightarrow \quad (x_n - 2)(x_n + 1) < 0, \end{aligned}$$

which holds true since $x_n > 0$, and since, as shown above, $x_n < 2$ for all $n \geq 0$.

Thus, the sequence $(x_n)_{n\geq 0}$ is convergent since it is increasing and bounded from above, which is what we needed to show in order to prove (6.6). □

Question 7. Find x such that

$$x^{x^{x^{\cdot^{\cdot^{\cdot}}}}} = 2. \tag{6.7}$$

Answer: If x exists such that (6.7) holds true, then

$$x^{x^{x^{x^{\cdot^{\cdot^{\cdot}}}}}} = x^2 = 2,$$

and therefore the only possible solution to (6.7) is $x = \sqrt{2}$. We prove that $x = \sqrt{2}$ is, indeed, the solution to (6.7), by showing that

$$\sqrt{2}^{\sqrt{2}^{\sqrt{2}^{\cdot^{\cdot^{\cdot}}}}} = 2. \tag{6.8}$$

Consider the sequence $(x_n)_{n\geq 0}$ with $x_0 = \sqrt{2}$ and satisfying the following recursion:

$$x_{n+1} = \left(\sqrt{2}\right)^{x_n} = 2^{x_n/2}, \ \forall\, n \geq 0. \tag{6.9}$$

We can see by induction that $(x_n)_{n\geq 0}$ is an increasing sequence, since $x_0 = \sqrt{2} < \sqrt{2}^{\sqrt{2}} = x_1$, and, if we assume that $x_{n-1} < x_n$, then

$$x_n = 2^{x_{n-1}/2} < 2^{x_n/2} = x_{n+1}.$$

Also by induction, we can prove that the sequence $(x_n)_{n\geq 0}$ is bounded from above by 2, since $x_0 = \sqrt{2} < 2$, and, if $x_n < 2$, then

$$x_{n+1} = 2^{x_n/2} < 2.$$

Thus, the sequence $(x_n)_{n\geq 0}$ is convergent since it is increasing and bounded from above.

Let $l = \lim_{n\to\infty} x_n$. From (6.9), we find that $l = 2^{l/2}$, which is equivalent to

$$l^{1/l} = 2^{1/2}. \tag{6.10}$$

The function $f : (0,\infty) \to (0,\infty)$ given by

$$\begin{aligned} f(t) &= t^{1/t} = \exp\left(\ln(t^{1/t})\right) \\ &= \exp\left(\frac{\ln(t)}{t}\right) \end{aligned}$$

is increasing for $t < e$ and decreasing for $t > e$, since

$$f'(t) = \frac{1 - \ln(t)}{t^2} \exp\left(\frac{\ln(t)}{t}\right)$$

and $f'(t) > 0$ for $t < e$ and $f'(t) < 0$ for $t > e$.

Thus, there are two values of l such that (6.10) is satisfied, i.e., such that $l^{1/l} = 2^{1/2}$, one value being equal to 2, and the other one greater than e. Since we showed that $x_n < 2$ for all $n \geq 0$, we conclude that $l = \lim_{n\to\infty} x_n = 2$, which is what we needed to show to complete the proof of (6.8). $\square$

Question 8. Which of the following series converge:

$$\sum_{k=1}^{\infty} \frac{1}{k}; \quad \sum_{k=1}^{\infty} \frac{1}{k^2}; \quad \sum_{k=2}^{\infty} \frac{1}{k\ln(k)}?$$

Answer: We show that

$$\sum_{k=1}^{\infty} \frac{1}{k^2} \quad \text{is convergent;}$$

$$\sum_{k=1}^{\infty} \frac{1}{k} \quad \text{and} \quad \sum_{k=2}^{\infty} \frac{1}{k\ln(k)} \quad \text{are divergent.}$$

Since all the terms of the series $\sum_{k=1}^{\infty} \frac{1}{k^2}$ are positive, it is enough to show that the partial sums $\sum_{k=1}^{n} \frac{1}{k^2}$ are uniformly bounded, in order to conclude that the series is convergent. This can be seen as follows:

$$\begin{aligned}
\sum_{k=1}^{n} \frac{1}{k^2} &= 1 + \sum_{k=2}^{n} \frac{1}{k^2} \\
&\leq 1 + \sum_{k=2}^{n} \frac{1}{k(k-1)} \\
&= 1 + \sum_{k=2}^{n} \left(\frac{1}{k-1} - \frac{1}{k} \right) \\
&= 1 + \left(1 - \frac{1}{n} \right) \\
&< 2, \quad \forall\ n \geq 2.
\end{aligned}$$

To show that the series $\sum_{k=1}^{\infty} \frac{1}{k}$ is divergent, we will prove that

$$\sum_{k=1}^{n} \frac{1}{k} > \ln(n) + \frac{1}{n}, \quad \forall\, n \geq 1. \tag{6.11}$$

Since $\frac{1}{x}$ is a decreasing function, it follows that

$$\frac{1}{x} < \frac{1}{k}, \quad \forall\, k < x < k+1.$$

Then,

$$\begin{aligned}
\int_1^n \frac{1}{x}\, dx &= \sum_{k=1}^{n-1} \int_k^{k+1} \frac{1}{x}\, dx \\
&< \sum_{k=1}^{n-1} \int_k^{k+1} \frac{1}{k}\, dx \\
&= \sum_{k=1}^{n-1} \frac{1}{k} \\
&= -\frac{1}{n} + \sum_{k=1}^{n} \frac{1}{k}.
\end{aligned} \tag{6.12}$$

From (6.12), we find that

$$\begin{aligned}
\sum_{k=1}^{n} \frac{1}{k} &> \int_1^n \frac{1}{x}\, dx + \frac{1}{n} \\
&= \ln(n) + \frac{1}{n}, \quad \forall\, n \geq 1,
\end{aligned}$$

which is what we wanted to prove; see (6.11).

Similarly, note that

$$\frac{1}{x\ln(x)} < \frac{1}{k\ln(k)}, \quad \forall\, k < x < k+1,$$

and therefore

$$\begin{aligned}\int_2^{n+1} \frac{1}{x\ln(x)}\,dx &= \sum_{k=2}^{n} \int_k^{k+1} \frac{1}{x\ln(x)}\,dx \\ &< \sum_{k=2}^{n} \int_k^{k+1} \frac{1}{k\ln(k)}\,dx \\ &= \sum_{k=2}^{n} \frac{1}{k\ln(k)}. \qquad (6.13)\end{aligned}$$

Since

$$\int_2^{n+1} \frac{1}{x\ln(x)}\,dx = \ln(\ln(n+1)) - \ln(\ln(2)),$$

we obtain from (6.13) that

$$\sum_{k=2}^{n} \frac{1}{k\ln(k)} > \ln(\ln(n+1)) - \ln(\ln(2)),$$

and we conclude that the series $\sum_{k=2}^{\infty} \frac{1}{k\ln(k)}$ is divergent. Note: Although not needed to answer this question, it can be shown that

$$\sum_{k=1}^{\infty} \frac{1}{k^2} = \frac{\pi^2}{6},$$

and

$$\lim_{n\to\infty} \left(\sum_{k=1}^{\infty} \frac{1}{k} - \ln(n) \right) = \gamma,$$

where $\gamma \approx 0.57721$ is Euler's constant. □

Question 9. Compute

$$\int \frac{1}{1+x^2}\,dx.$$

Answer: Use the substitution $x = \tan(z)$. Then, $dx = \frac{1}{\cos^2(z)} dz$ and

$$\begin{aligned}\int \frac{1}{1+x^2}\,dx &= \int \frac{1}{(1+\tan^2(z))\cos^2(z)}\,dz \\ &= \int \frac{1}{\cos^2(z)+\sin^2(z)}\,dz \\ &= \int 1\,dz \\ &= z + C,\end{aligned}$$

where C is a real constant, since $\cos^2(z) + \sin^2(z) = 1$ for any z. Solving $x = \tan(z)$ for z, we obtain that $z = \arctan(x)$, and therefore

$$\int \frac{1}{1+x^2}\,dx = \arctan(x) + C. \quad \square$$

Question 10. Compute

$$\int x\ln(x)dx \quad \text{and} \quad \int xe^x\,dx.$$

Answer: By integration by parts,

$$\begin{aligned}\int x\ln(x)\,dx &= \frac{x^2}{2}\ln(x) - \int \frac{x^2}{2}\cdot\frac{1}{x}\,dx \\ &= \frac{x^2}{2}\ln(x) - \frac{1}{2}\int x\,dx \\ &= \frac{x^2\ln(x)}{2} - \frac{x^2}{4} + C; \\ \int xe^x\,dx &= xe^x - \int 1\cdot e^x\,dx \\ &= xe^x - e^x + C. \quad \square\end{aligned}$$

Question 11. Compute

$$\int x^n \ln(x)\, dx.$$

Answer: If $n \neq -1$, we use integration by parts and find that

$$\begin{aligned}\int x^n \ln(x)\, dx &= \frac{x^{n+1}}{n+1}\ln(x) - \int \frac{x^{n+1}}{n+1}\cdot\frac{1}{x}\, dx \\ &= \frac{x^{n+1}\ln(x)}{n+1} - \frac{1}{n+1}\int x^n\, dx \\ &= \frac{x^{n+1}\ln(x)}{n+1} - \frac{x^{n+1}}{(n+1)^2} + C.\end{aligned}$$

For $n = -1$, we obtain that

$$\int \frac{\ln(x)}{x}\, dx = \frac{(\ln(x))^2}{2} + C,$$

where C is a real constant, since

$$\left(\frac{(\ln(x))^2}{2}\right)' = \frac{1}{2}\cdot 2\ln(x)\cdot(\ln(x))' = \frac{\ln(x)}{x}. \quad \square$$

Question 12. Compute

$$\int (\ln(x))^n\, dx.$$

Answer: For every integer $n \geq 0$, let

$$f_n(x) = \int (\ln(x))^n\, dx.$$

By using integration by parts, we find that, for any $n \geq 1$,

$$\begin{aligned}
& \int (\ln(x))^n \, dx \\
= & \; x(\ln(x))^n - \int x \left((\ln(x))^n\right)' \, dx \\
= & \; x(\ln(x))^n - \int x \cdot n(\ln(x))^{n-1} \cdot (\ln(x))' \, dx \\
= & \; x(\ln(x))^n - \int x \cdot n(\ln(x))^{n-1} \cdot \frac{1}{x} \, dx \\
= & \; x(\ln(x))^n - n \int (\ln(x))^{n-1} \, dx,
\end{aligned}$$

and therefore

$$f_n(x) = x(\ln(x))^n - n f_{n-1}(x), \;\; \forall \, n \geq 1. \tag{6.14}$$

Note that

$$f_0(x) = \int 1 dx = x + C.$$

Thus, the recursion (6.14) can be used to find the values of $f_n(x)$ for all n. For example:

$$\begin{aligned}
f_1(x) &= x \ln(x) - f_0(x) = x(\ln(x) - 1) + C; \\
f_2(x) &= x(\ln(x))^2 - 2 f_1(x) \\
&= x \left((\ln(x))^2 - 2\ln(x) + 2\right) + C.
\end{aligned}$$

The following general formula can be obtained by induction:

$$\int (\ln(x))^n \, dx = x \sum_{k=0}^{n} \frac{(-1)^{n-k} n!}{k!} (\ln(x))^k + C,$$

for all $n \geq 0$. □

Question 13. Solve the ODE

$$y'' - 4y' + 4y = 1. \tag{6.15}$$

Answer: Note that (6.15) is a second order non homogeneous linear ODE with constant coefficients. The homogeneous ODE associated to (6.15) is

$$y'' - 4y' + 4y = 0, \tag{6.16}$$

whose characteristic equation, $z^2 - 4z + 4 = 0$, has a double root $z_1 = z_2 = 2$. Thus, the solution to the homogeneous ODE (6.16) is

$$y(x) = c_1 e^{2x} + c_2 x e^{2x}, \tag{6.17}$$

where c_1 and c_2 are constants.

Since the constant function $y_0(x) = \frac{1}{4}$ is a solution to the non homogeneous ODE (6.15), we conclude from (6.17) that the general form of the solution of (6.15) is

$$y(x) = c_1 e^{2x} + c_2 x e^{2x} + \frac{1}{4}. \quad \square$$

Question 14. Find $f(x)$ such that

$$f'(x) = f(x)(1 - f(x)). \tag{6.18}$$

Answer: Note that (6.18) is as an ODE with separable variables and can be written as follows:

$$\frac{y'}{y(1-y)} = 1, \tag{6.19}$$

where $y = f(x)$. By integrating (6.19) with respect to x we obtain that

$$\int \frac{y'}{y(1-y)}\, dx = \int 1\, dx = x + C_1, \tag{6.20}$$

where $C_1 \in \mathbb{R}$ is a real constant.

Note that $dy = y'dx$, and therefore

$$\begin{aligned}
\int \frac{y'}{y(1-y)}\,dx &= \int \frac{1}{y(1-y)}\,dy \\
&= \int \left(\frac{1}{y} + \frac{1}{1-y} \right) dy \\
&= \int \frac{1}{y}\,dy + \int \frac{1}{1-y}\,dy \\
&= \ln(|y|) - \ln(|1-y|) \\
&= \ln\left| \frac{y}{1-y} \right|. \qquad (6.21)
\end{aligned}$$

From (6.20) and (6.21), it follows that

$$\ln\left| \frac{y}{1-y} \right| = x + C_1,$$

and therefore

$$\left| \frac{y}{1-y} \right| = e^{x+C_1} = C_2 e^x,$$

where $C_2 = e^{C_1} > 0$ is a positive real constant.

Thus, either $\frac{y}{1-y} = C_2 e^x$, or $\frac{y}{1-y} = -C_2 e^x$, which can be written as

$$\frac{y}{1-y} = Ce^x, \qquad (6.22)$$

where C is a real constant.

From (6.22), we obtain that $y = \frac{Ce^x}{1+Ce^x}$. We conclude that the ODE (6.18) has the following solution:

$$f(x) = \frac{Ce^x}{1+Ce^x},$$

where $C \in \mathbb{R}$ is a fixed constant. □

Question 15. Derive the Black-Scholes PDE.

Answer: Consider an asset with spot price S following a lognormal distribution with drift μ and volatility σ and paying dividends continuously at rate q. Then,

$$dS = (\mu - q)Sdt + \sigma S dW_t,$$

where W_t, $t \geq 0$, is a Wiener process.

Let $V = V(S,t)$ be the value at time t of a replicable non path dependent derivative security on this asset, when the underlying is priced at S. Set up a portfolio Π made of a long position in the derivative security V and a short position in

$$\Delta = \frac{\partial V}{\partial S} \tag{6.23}$$

units of the asset. Then,

$$\Pi = V - \Delta S.$$

Denote by dS, dV, and $d\Pi$ the changes in the values of S, V, and Π, respectively, over an infinitesimally small time period dt. Then,

$$d\Pi = dV - \Delta dS - \Delta qSdt, \tag{6.24}$$

where $\Delta qSdt$ is the dividend payment owed over the time dt on the short Δ units asset position. From (6.23) and (6.24), we find that

$$d\Pi = dV - \frac{\partial V}{\partial S}dS - qS\frac{\partial V}{\partial S}dt. \tag{6.25}$$

From Itô's formula, it follows that

$$\begin{aligned} dV &= \left(\frac{\partial V}{\partial t} + \frac{\sigma^2 S^2}{2}\frac{\partial^2 V}{\partial S^2}\right)dt \\ &\quad + \frac{\partial V}{\partial S}dS, \end{aligned} \tag{6.26}$$

and, from (6.25) and (6.26), we obtain that

$$d\Pi = \left(\frac{\partial V}{\partial t} + \frac{\sigma^2 S^2}{2}\frac{\partial^2 V}{\partial S^2} - qS\frac{\partial V}{\partial S}\right)dt, \tag{6.27}$$

which means that the value of the portfolio Π is deterministic over the small time period dt. For no–arbitrage, the value of the portfolio Π must grow at the risk–free rate over the time period dt, i.e., $d\Pi = r\Pi dt$, where r denotes the risk–free rate. Thus,

$$\begin{aligned} d\Pi &= r\Pi dt = r(V - \Delta S)dt \\ &= \left(rV - rS\frac{\partial V}{\partial S}\right) dt. \end{aligned} \qquad (6.28)$$

From (6.27) and (6.28), we find that

$$\frac{\partial V}{\partial t} + \frac{\sigma^2 S^2}{2}\frac{\partial^2 V}{\partial S^2} - qS\frac{\partial V}{\partial S} = rV - rS\frac{\partial V}{\partial S},$$

and therefore

$$\frac{\partial V}{\partial t} + \frac{\sigma^2 S^2}{2}\frac{\partial^2 V}{\partial S^2} + (r-q)S\frac{\partial V}{\partial S} - rV = 0,$$

which is the Black-Scholes PDE for $V(S,t)$. □

6.2 Covariance and correlation matrices. Linear algebra.

Question 1. Show that any covariance matrix is symmetric positive semidefinite. Show that the same is true for correlation matrices.

Answer: Let Σ_X and Ω_X be the covariance matrix and the correlation matrix of n random variables X_1, X_2, ..., X_n. It is easy to see that Σ_X and Ω_X are symmetric matrices:

$$\begin{aligned}\Sigma_X(j,k) &= \operatorname{cov}(X_j, X_k) = \operatorname{cov}(X_k, X_j) \\ &= \Sigma_X(k,j), \ \forall\, 1 \le j,k \le n; \\ \Omega_X(j,k) &= \operatorname{corr}(X_j, X_k) = \operatorname{corr}(X_k, X_j) \\ &= \Omega_X(k,j), \ \forall\, 1 \le j,k \le n.\end{aligned}$$

Let c_1, c_2, ..., c_n be real numbers, and let $C = (c_i)_{i=1:n}$ be a column vector of size n. Recall that

$$\operatorname{var}\left(\sum_{i=1}^n c_i X_i\right) = C^t \Sigma_X C. \qquad (6.29)$$

Since the variance of any random variable is nonnegative, it follows that

$$C^t \Sigma_X C \ge 0, \ \forall\, C \in \mathbb{R}^n, \qquad (6.30)$$

and we conclude that Σ_X is a symmetric positive semidefinite matrix.

For completeness, we include a proof of (6.29) here. Let $Y = \sum_{i=1}^n c_i X_i$. Then,

$$Y - E[Y] = \sum_{i=1}^n c_i (X_i - \mu_i),$$

where $\mu_i = E[X_i]$, for $i = 1:n$, and therefore

$$\begin{aligned}
& \operatorname{var}\left(\sum_{i=1}^{n} c_i X_i\right) \\
= \ & \operatorname{var}(Y) \ = \ E\left[(Y - E[Y])^2\right] \\
= \ & E\left[\left(\sum_{i=1}^{n} c_i (X_i - \mu_i)\right)^2\right] \qquad (6.31) \\
= \ & E\left[\sum_{1 \le j,k \le n} c_j c_k (X_j - \mu_j)(X_k - \mu_k)\right] \\
= \ & \sum_{1 \le j,k \le n} c_j c_k E[(X_j - \mu_j)(X_k - \mu_k)] \\
= \ & \sum_{1 \le j,k \le n} c_j c_k \operatorname{cov}(X_j, X_k) \\
= \ & \sum_{1 \le j,k \le n} c_j c_k \Sigma_X(j,k) \\
= \ & C^t \Sigma_X C, \qquad (6.32)
\end{aligned}$$

where, for (6.31) and for (6.32), we used the following facts, respectively:

$$\begin{aligned}
\left(\sum_{i=1}^{n} z_i\right)^2 \ &= \ \sum_{1 \le j,k \le n} z_j z_k, \ \ \forall \ z_i \in \mathbb{R}, \ i = 1:n; \\
x^t A x \ &= \ \sum_{1 \le j,k \le n} x_j x_k A(j,k),
\end{aligned}$$

for any $n \times 1$ vector $x = (x_i)_{i=1:n} \in \mathbb{R}^n$, and for any $n \times n$ matrix A.

To show that Ω_X is a symmetric positive semidefinite matrix, recall the following correspondence between covariance matrices and correlation matrices:

$$\Sigma_X \ = \ D_{\sigma_X} \Omega_X D_{\sigma_X}, \qquad (6.33)$$

where $D_{\sigma_X} = \mathrm{diag}(\sigma_i)_{i=1:n}$ is the diagonal matrix with entries equal to the standard deviations of the n random variables, i.e., $\sigma_i^2 = \mathrm{var}(X_i)$, for $i = 1 : n$.

Note that $(D_{\sigma_X})^{-1} = \mathrm{diag}\left(\frac{1}{\sigma_i}\right)_{i=1:n}$. Let $v \in \mathbb{R}^n$, and let

$$w = (D_{\sigma_X})^{-1} v.$$

Then,

$$w^t = v^t \left((D_{\sigma_X})^{-1}\right)^t = v^t (D_{\sigma_X})^{-1}, \qquad (6.34)$$

since $(D_{\sigma_X})^{-1}$ is a diagonal matrix and therefore symmetric, i.e., $\left((D_{\sigma_X})^{-1}\right)^t = (D_{\sigma_X})^{-1}$.

From (6.33), (6.34), and (6.30), we find that

$$\begin{aligned} w^t \Sigma_X w &= w^t (D_{\sigma_X} \Omega_X D_{\sigma_X}) w \\ &= v^t (D_{\sigma_X})^{-1} (D_{\sigma_X} \Omega_X D_{\sigma_X}) (D_{\sigma_X})^{-1} v \\ &= v^t \Omega_X v \\ &\geq 0. \end{aligned}$$

Thus, $v^t \Omega_X v \geq 0$ for all $v \in \mathbb{R}^n$, and we conclude that Ω_X is a symmetric positive semidefinite matrix. □

Question 2. Find the correlation matrix of three random variables with covariance matrix

$$\Sigma_X = \begin{pmatrix} 1 & 0.36 & -1.44 \\ 0.36 & 4 & 0.80 \\ -1.44 & 0.80 & 9 \end{pmatrix}. \qquad (6.35)$$

Answer: If Ω_X is the correlation matrix of the three random variables, then

$$\Sigma_X = \begin{pmatrix} \sigma_1 & 0 & 0 \\ 0 & \sigma_2 & 0 \\ 0 & 0 & \sigma_3 \end{pmatrix} \Omega_X \begin{pmatrix} \sigma_1 & 0 & 0 \\ 0 & \sigma_2 & 0 \\ 0 & 0 & \sigma_3 \end{pmatrix},$$

where σ_i, $i = 1 : 3$, are the standard deviations of the three random variables, and therefore

$$\Omega_X = \begin{pmatrix} \frac{1}{\sigma_1} & 0 & 0 \\ 0 & \frac{1}{\sigma_2} & 0 \\ 0 & 0 & \frac{1}{\sigma_3} \end{pmatrix} \Sigma_X \begin{pmatrix} \frac{1}{\sigma_1} & 0 & 0 \\ 0 & \frac{1}{\sigma_2} & 0 \\ 0 & 0 & \frac{1}{\sigma_3} \end{pmatrix}.$$

Since the standard deviations of the random variables with covariance matrix Σ_X given by (6.35) are

$$\begin{aligned} \sigma_1 &= \sqrt{\Sigma_X(1,1)} = 1; \\ \sigma_2 &= \sqrt{\Sigma_X(2,2)} = 2; \\ \sigma_3 &= \sqrt{\Sigma_X(2,2)} = 3, \end{aligned}$$

we obtain that

$$\begin{aligned} \Omega_X &= \begin{pmatrix} 1 & 0 & 0 \\ 0 & \frac{1}{2} & 0 \\ 0 & 0 & \frac{1}{3} \end{pmatrix} \Sigma_X \begin{pmatrix} 1 & 0 & 0 \\ 0 & \frac{1}{2} & 0 \\ 0 & 0 & \frac{1}{3} \end{pmatrix} \\ &= \begin{pmatrix} 1 & 0.18 & -0.48 \\ 0.18 & 1 & 0.1333 \\ -0.48 & 0.1333 & 1 \end{pmatrix}. \quad \square \end{aligned}$$

Question 3. Assume that all the entries of an $n \times n$ correlation matrix which are not on the main diagonal are equal to ρ. Find upper and lower bounds on the possible values of ρ.

Answer: Recall that a symmetric matrix with diagonal entries equal to 1 is a correlation matrix if and only if the matrix is symmetric positive semidefinite, i.e., if and only if all the eigenvalues of the matrix are nonnegative.

Let

$$\Omega = \begin{pmatrix} 1 & \rho & \dots & \rho \\ \rho & \ddots & \ddots & \vdots \\ \vdots & \ddots & \ddots & \rho \\ \rho & \dots & \rho & 1 \end{pmatrix}.$$

We include two ways to compute the eigenvalues of Ω, which are then used to find the necessary and sufficient conditions for the matrix Ω to be a correlation matrix.

Solution 1: Note that

$$\begin{aligned} \Omega &= (1-\rho)I + \begin{pmatrix} \rho & \rho & \dots & \rho \\ \rho & \ddots & \ddots & \vdots \\ \vdots & \ddots & \ddots & \rho \\ \rho & \dots & \rho & \rho \end{pmatrix} \\ &= (1-\rho)I + \rho M, \end{aligned}$$

where M is the $n \times n$ matrix with all entries equal to 1, and I is the $n \times n$ identity matrix.

Let λ and $v = (v_i)_{i=1:n}$ be an eigenvalue and a corresponding eigenvector of M, i.e., $Mv = \lambda v$, with $v \neq 0$.

Then, $Mv = \lambda v$ can be written as

$$\begin{cases} v_1 + v_2 + \cdots + v_n &= \lambda v_1; \\ v_1 + v_2 + \cdots + v_n &= \lambda v_2; \\ \vdots & \vdots \\ v_1 + v_2 + \cdots + v_n &= \lambda v_n, \end{cases}$$

and therefore

$$\lambda v_1 = \lambda v_2 = \ldots = \lambda v_n$$

Thus, either $\lambda = 0$, or $v_1 = v_2 = \cdots = v_n$, in which case $nv_1 = \lambda v_1$, and therefore $\lambda = n$, since $v = (v_i)_{i=1:n} \neq 0$.

In other words, the eigenvalues of M are $\lambda = 0$ and $\lambda = n$.

Note that, if $Mv = \lambda v$, then,

$$\begin{aligned}\Omega v &= (1-\rho)v + \rho M v \;=\; (1-\rho)v + \rho\lambda v \\ &= (1-\rho+\rho\lambda)v.\end{aligned}$$

Thus, $\mu = 1 - \rho + \rho\lambda$ and v are an eigenvalue and corresponding eigenvector of Ω.

Since the eigenvalues of M are $\lambda = 0$ and $\lambda = n$, it follows that the eigenvalues[1] of Ω are $\mu = 1 - \rho$, corresponding to $\lambda = 0$, and $\mu = (1-\rho) + n\rho = 1 + (n-1)\rho$, corresponding to $\lambda = n$.

Since Ω is a correlation matrix if and only if all its eigenvalues are nonnegative, we conclude that the matrix Ω is a correlation matrix if and only if

$$0 \leq 1 + (n-1)\rho \quad \text{and} \quad 0 \leq 1 - \rho,$$

which is equivalent to

$$-\frac{1}{n-1} \;\leq\; \rho \;\leq\; 1. \tag{6.36}$$

[1] The eigenvalue $1-\rho$ has multiplicity $n-1$, and the eigenvalue $1+(n-1)\rho$ has multiplicity 1; see Solution 2 of this question.

Solution 2: Note that

$$\begin{aligned}
\Omega &= (1-\rho)I + \rho \begin{pmatrix} 1 & 1 & \dots & 1 \\ 1 & \ddots & \ddots & \vdots \\ \vdots & \ddots & \ddots & 1 \\ 1 & \dots & 1 & 1 \end{pmatrix} \\
&= (1-\rho)I + \rho \begin{pmatrix} 1 \\ \vdots \\ 1 \end{pmatrix} (1 \ \dots \ 1) \\
&= (1-\rho)I + \rho w w^t \\
&= (1-\rho)I + \rho A,
\end{aligned}$$

where I is the $n \times n$ identity matrix and A is the $n \times n$ matrix given by $A = ww^t$, where w is the $n \times 1$ column vector of size n with all entries equal to 1.

Recall that an $n \times n$ matrix of the form uu^t, where $u = (u_i)_{i=1:n}$ is an $n \times 1$ column vector, has an eigenvalue equal to $\sum_{i=1}^n u_i^2$ with multiplicity 1 and another eigenvalue equal to 0 with multiplicity $n-1$.

Then, the eigenvalues of the matrix A are:
$\lambda = \sum_{i=1}^n w_i^2 = \sum_{i=1}^n 1 = n$ with multiplicity 1;
$\lambda = 0$ with multiplicity $n-1$.

Note that, if λ and v are an eigenvalue and a corresponding eigenvector of A, then $Av = \lambda v$, and therefore

$$\begin{aligned}
\Omega v &= (1-\rho)v + \rho A v = (1-\rho)v + \rho\lambda v \\
&= (1-\rho+\rho\lambda)v.
\end{aligned}$$

Thus, $1-\rho+\rho\lambda$ and v are an eigenvalue and corresponding eigenvector of Ω. and we obtain that the matrix Ω has the following eigenvalues:

- $(1-\rho) + n\rho = 1 + (n-1)\rho$ with multiplicity 1;
- $1-\rho$ with multiplicity $n-1$.

As before, since Ω is a correlation matrix if and only if all its eigenvalues are nonnegative, we conclude that the matrix Ω is a correlation matrix if and only if

$$0 \leq 1 + (n-1)\rho \quad \text{and} \quad 0 \leq 1 - \rho,$$

which is equivalent to

$$\frac{1}{n-1} \leq \rho \leq 1,$$

which is the same as (6.36). $\square$

Question 4. How many eigenvalues does an $n \times n$ matrix with real entries have? How many eigenvectors?

Answer: Any $n \times n$ matrix with real entries has n eigenvalues, counted with their multiplicities; some of the eigenvalues may be complex numbers. Any $n \times n$ matrix has at most n eigenvectors.

Let A be an $n \times n$ matrix. Let λ be an eigenvalue of A with corresponding eigenvector $v \neq 0$, and let $P_A(x) = \det(xI_n - A)$ be the characteristic polynomial of A, where I_n is the $n \times n$ identity matrix. Note that

$$\begin{aligned} Av = \lambda v, \; v \neq 0 \quad &\Longleftrightarrow \quad (\lambda I_n - A)v = 0, \; v \neq 0 \\ &\Longleftrightarrow \quad \lambda I_n - A \text{ singular matrix} \\ &\Longleftrightarrow \quad \det(\lambda I_n - A) = 0 \\ &\Longleftrightarrow \quad P_A(\lambda) = 0. \end{aligned}$$

In other words, λ is an eigenvalue of A if and only if λ is a root of the corresponding characteristic polynomial $P_A(x)$. Since $P_A(x)$ is a polynomial of degree n, it follows from the Fundamental Theorem of Algebra that $P_A(x)$ has exactly n (complex) roots when counted with their multiplicities. We conclude that any $n \times n$ matrix has n eigenvalues, counted with their multiplicities.

An eigenvalue of multiplicity m has at least one eigenvector and at most m linearly independent corresponding eigenvectors, but it may have less than m linearly independent eigenvectors.[2] Thus, an $n \times n$ matrix has at most n eigenvectors, and at least as many eigenvectors as the number of distinct eigenvalues of the matrix. □

Question 5. Let

$$A = \begin{pmatrix} 2 & -2 \\ -2 & 5 \end{pmatrix}.$$

(i) Find a 2×2 matrix M such that $M^2 = A$;

(ii) Find a 2×2 matrix M such that $A = MM^t$.

Answer: (i) Recall that any symmetric matrix has the diagonal form

$$A = Q\Lambda Q^t, \tag{6.37}$$

where Λ is the diagonal matrix whose entries on the main diagonal are the eigenvalues of A and Q is the orthogonal matrix whose columns are the corresponding eigenvectors of A of norm 1, i.e.,

$$\Lambda = \begin{pmatrix} \lambda_1 & 0 \\ 0 & \lambda_2 \end{pmatrix}; \quad Q = (v_1 \ v_2), \tag{6.38}$$

where $Av_1 = \lambda_1 v_1$ and $Av_2 = \lambda_2 v_2$, with $||v_1|| = ||v_2|| = 1$.

If the matrix A has nonnegative eigenvalues, i.e., if $\lambda_1 \geq 0$ and $\lambda_2 \geq 0$, then the matrix

$$M = Q\Lambda^{1/2}Q^t \tag{6.39}$$

[2]For example, the matrix $\begin{pmatrix} 2 & 1 & 0 \\ 0 & 2 & 1 \\ 0 & 0 & 2 \end{pmatrix}$ has eigenvalue 2 with multiplicity 3 and only one eigenvector, $\begin{pmatrix} 1 \\ 0 \\ 0 \end{pmatrix}$.

with

$$\Lambda^{1/2} = \begin{pmatrix} \sqrt{\lambda_1} & 0 \\ 0 & \sqrt{\lambda_2} \end{pmatrix} \tag{6.40}$$

has the property that $M^2 = A$:

$$\begin{aligned} M^2 &= \left(Q\Lambda^{1/2}Q^t\right)\left(Q\Lambda^{1/2}Q^t\right) \\ &= Q\Lambda^{1/2}(Q^tQ)\Lambda^{1/2}Q^t \\ &= Q\Lambda^{1/2}\Lambda^{1/2}Q^t \\ &= Q\Lambda Q^t \\ &= A, \end{aligned}$$

since Q is an orthogonal matrix and therefore $Q^tQ = I$, and since, from (6.40), it follows that $\Lambda^{1/2}\Lambda^{1/2} = \Lambda$.

We now proceed to compute the eigenvalues and the eigenvectors of the matrix A. The eigenvalues of the matrix A are the roots of the characteristic polynomial $P_A(x)$ of the matrix A given by[3]

$$\begin{aligned} P_A(x) &= \det(xI - A) = \det\begin{pmatrix} x-2 & 2 \\ 2 & x-5 \end{pmatrix} \\ &= (x-2)(x-5) - 4 = x^2 - 7x + 6 \\ &= (x-1)(x-6). \end{aligned}$$

The roots of $P_A(x)$ are 1 and 6, and therefore the eigenvalues of A are $\lambda_1 = 1$ and $\lambda_2 = 6$. The corresponding eigenvectors of norm 1 are

$$v_1 = \begin{pmatrix} \frac{2}{\sqrt{5}} \\ \frac{1}{\sqrt{5}} \end{pmatrix} \quad \text{and} \quad v_2 = \begin{pmatrix} \frac{1}{\sqrt{5}} \\ -\frac{2}{\sqrt{5}} \end{pmatrix}.$$

[3] The characteristic polynomial of the matrix A can also be obtained as follows:

$$P_A(x) = x^2 - \operatorname{tr}(A)x + \det(A) = x^2 - 7x + 6,$$

where $\operatorname{tr}(A) = 2 + 5 = 7$ and $\det(A) = 2 \cdot 5 - (-2) \cdot (-2) = 6$.

For example, if $\lambda_2 = 6$, any corresponding eigenvector $v_2 = \begin{pmatrix} a \\ b \end{pmatrix} \neq 0$ is a solution to $Av = 6v$, which can be written as

$$\left\{ \begin{array}{rcl} 2a - 2b & = & 6a \\ -2a + 5b & = & 6b \end{array} \right. \iff \left\{ \begin{array}{rcl} b & = & -2a \\ b & = & -2a \end{array} \right.$$

Thus, any eigenvector corresponding to the eigenvalue $\lambda_2 = 6$ is of the form

$$v_2 = \begin{pmatrix} a \\ -2a \end{pmatrix} = a \begin{pmatrix} 1 \\ -2 \end{pmatrix}.$$

By choosing $a = \frac{1}{\sqrt{5}}$, we obtain that an eigenvector of norm 1 corresponding to the eigenvalue $\lambda_2 = 6$ is

$$v_2 = \begin{pmatrix} \frac{1}{\sqrt{5}} \\ -\frac{2}{\sqrt{5}} \end{pmatrix}.$$

Then, it follows from (6.39) and (6.40) that the matrix M given by

$$\begin{aligned} M &= \begin{pmatrix} \frac{2}{\sqrt{5}} & \frac{1}{\sqrt{5}} \\ \frac{1}{\sqrt{5}} & -\frac{2}{\sqrt{5}} \end{pmatrix} \begin{pmatrix} \sqrt{1} & 0 \\ 0 & \sqrt{6} \end{pmatrix} \begin{pmatrix} \frac{2}{\sqrt{5}} & \frac{1}{\sqrt{5}} \\ \frac{1}{\sqrt{5}} & -\frac{2}{\sqrt{5}} \end{pmatrix} \\ &= \frac{1}{5} \begin{pmatrix} 2 & 1 \\ 1 & -2 \end{pmatrix} \begin{pmatrix} 1 & 0 \\ 0 & \sqrt{6} \end{pmatrix} \begin{pmatrix} 2 & 1 \\ 1 & -2 \end{pmatrix} \\ &= \frac{1}{5} \begin{pmatrix} 4 + \sqrt{6} & 2 - 2\sqrt{6} \\ 2 - 2\sqrt{6} & 1 + 4\sqrt{6} \end{pmatrix} \end{aligned}$$

has the property that $M^2 = A$.

(ii) We found that the eigenvalues of A are 1 and 6, i.e., positive. Then, A is a symmetric positive definite matrix, and therefore has a Cholesky decomposition. Recall that, if U is the Cholesky factor of the matrix A, then, by definition, $A = U^t U$. Thus, in order to find a matrix M such that $A = MM^t$, it is enough to compute the Cholesky factor

$$U = \begin{pmatrix} U(1,1) & U(1,2) \\ 0 & U(2,2) \end{pmatrix}$$

of the matrix A and let $M = U^t$.

Note that $A = U^t U$ can be written as

$$\begin{pmatrix} A(1,1) & A(1,2) \\ A(2,1) & A(2,2) \end{pmatrix} \tag{6.41}$$
$$= \begin{pmatrix} U(1,1) & 0 \\ U(1,2) & U(2,2) \end{pmatrix} \begin{pmatrix} U(1,1) & U(1,2) \\ 0 & U(2,2) \end{pmatrix}.$$

In the first step of the Cholesky decomposition, the first row of U is computed as follows:

$$\begin{aligned} U(1,1) &= \sqrt{A(1,1)} = \sqrt{2}; \\ U(1,2) &= \frac{A(1,2)}{U(1,1)} = \frac{-2}{\sqrt{2}} = -\sqrt{2}; \end{aligned}$$

see also (6.41).

In the next step of the Cholesky decomposition, the entry $A(2,2)$ of A is updated to $A(2,2) - U(1,2)^2 = 3$, and therefore

$$\begin{aligned} U(2,2) &= \sqrt{A(2,2) - (U(1,2))^2} \\ &= \sqrt{5 - (-\sqrt{2})^2} \\ &= \sqrt{3} \end{aligned}$$

Thus, the Cholesky factor of the matrix A is

$$U = \begin{pmatrix} \sqrt{2} & -\sqrt{2} \\ 0 & \sqrt{3} \end{pmatrix},$$

and therefore the matrix $M = U^t$ given by

$$M = \begin{pmatrix} \sqrt{2} & 0 \\ -\sqrt{2} & \sqrt{3} \end{pmatrix},$$

has the property that $A = MM^t$. $\square$

Question 6. The 2×2 matrix A has eigenvalues 2 and -3 with corresponding eigenvectors $\begin{pmatrix} 1 \\ 2 \end{pmatrix}$ and $\begin{pmatrix} -1 \\ 3 \end{pmatrix}$. If $v = \begin{pmatrix} 3 \\ 1 \end{pmatrix}$, find Av.

Answer: Let $\lambda_1 = 2$, $v_1 = \begin{pmatrix} 1 \\ 2 \end{pmatrix}$, and $\lambda_2 = -3$, $v_2 = \begin{pmatrix} -1 \\ 3 \end{pmatrix}$. We first find constants $c_1, c_2 \in \mathbb{R}$ such that $v = c_1 v_1 + c_2 v_2$, i.e., such that

$$\begin{cases} 3 = c_1 - c_2 \\ 1 = 2c_1 + 3c_2 \end{cases}$$

The solution of this linear system is $c_1 = 2$ and $c_2 = -1$. Thus,

$$v = 2v_1 - v_2. \tag{6.42}$$

Since $Av_1 = \lambda_1 v_1 = 2v_1$ and $Av_2 = \lambda_2 v_2 = -3v_2$, we find from (6.42) that

$$\begin{aligned} Av &= 2Av_1 - Av_2 = 2(2v_1) - (-3v_2) \\ &= 4v_1 + 3v_2 \\ &= \begin{pmatrix} 1 \\ 17 \end{pmatrix}. \quad \square \end{aligned}$$

Question 7. Let A and B be square matrices of the same size. Show that the traces of the matrices AB and BA are equal.

Answer: Recall that, for any two square matrices A and B of the same size, the matrices AB and BA have the same characteristic polynomial, i.e.,

$$\begin{aligned} P_{AB}(x) &= \det(xI - AB) = \det(xI - BA) \\ &= P_{BA}(x), \ \forall\, x \in \mathbb{R}, \end{aligned} \tag{6.43}$$

where I is the identity matrix of the same size as the matrices A and B.

Also, recall the following connection between the characteristic polynomial $P_M(x)$ of an $n \times n$ matrix M and the trace $\mathrm{tr}(M)$ of the matrix:

$$\begin{aligned} & P_M(x) \\ = \ & \det(xI - M) \\ = \ & x^n - \mathrm{tr}(M)x^{n-1} + \cdots + (-1)^n \det(M). \end{aligned} \tag{6.44}$$

Since $P_{AB}(x) = P_{BA}(x)$, see (6.43), we conclude from (6.44) that

$$\mathrm{tr}(AB) = \mathrm{tr}(BA). \tag{6.45}$$

For completeness, we include a proof of (6.43). If the matrix B is nonsingular, then

$$xI - AB = B^{-1}(xI - BA)B,$$

and therefore

$$\begin{aligned} \det(xI - AB) &= \det(B^{-1})\det(xI - BA)\det(B) \\ &= \det(xI - BA), \end{aligned} \tag{6.46}$$

since

$$\det(B^{-1})\det(B) = \det(B^{-1}B) = \det(I) = 1.$$

If the matrix B is singular, let ϵ be a real number, and note that the matrix $B - \epsilon I$ is singular if and only if ϵ is equal to an eigenvalue of B. Since the $n \times n$ matrix B has at most n eigenvalues, it follows that, except for a finite number of values of ϵ, the matrix $B - \epsilon I$ is nonsingular, in which case we obtain from (6.46) that

$$\det(xI - A(B - \epsilon I)) = \det(xI - (B - \epsilon I)A). \quad (6.47)$$

Since both sides of (6.47) are polynomials of degree n in ϵ, and therefore continuous functions of ϵ, we can let $\epsilon \to 0$ in (6.47) and obtain that

$$\begin{aligned} & \lim_{\epsilon \to 0} \left(\det(xI - A(B - \epsilon I)) \right) \\ & \quad = \lim_{\epsilon \to 0} \left(\det(xI - (B - \epsilon I)A) \right) \\ \Longleftrightarrow \quad & \det(xI - AB) = \det(xI - BA). \quad (6.48) \end{aligned}$$

From (6.46) and (6.48), we conclude that $\det(xI - AB) = \det(xI - BA)$ regardless of whether the matrix B is nonsingular or singular, which concludes the proof of (6.43). $\square$

Question 8. Can you find $n \times n$ matrices A and B such that

$$AB - BA = I_n,$$

where I_n is the identity matrix of size n?

Answer: We give a proof by contradiction. If it were possible to find $n \times n$ matrices A and B such that $AB - BA = I$, then

$$\operatorname{tr}(AB - BA) = \operatorname{tr}(I_n) = n. \quad (6.49)$$

However,

$$\operatorname{tr}(AB - BA) = \operatorname{tr}(AB) - \operatorname{tr}(BA) = 0, \quad (6.50)$$

since, if A and B are square matrices, then $\mathrm{tr}(AB) = \mathrm{tr}(BA)$; cf. (6.45).

Since (6.49) and (6.50) contradict each other, we conclude that there are no matrices A and B such that $AB - BA = I$. $\square$

Question 9. A probability matrix is a matrix with nonnegative entries such that the sum of the entries in each row of the matrix is equal to 1. Show that the product of two probability matrices is a probability matrix.

Answer: We first establish the following equivalent definition for a probability matrix:

The $n \times n$ matrix M is a probability matrix if and only if all the entries of M are nonnegative and

$$M\mathbf{1} = \mathbf{1}, \tag{6.51}$$

where $\mathbf{1}$ is the $n \times 1$ column vector with all entries equal to 1.

To see this, let $M = \mathrm{row}\,(r_j)_{j=1:n}$ be the row form of the matrix M, where r_j is an $1 \times n$ row vector, for $j = 1 : n$. The sum of all the entries in the j-th row r_j of M can be written as follows:[4]

$$\sum_{k=1}^{n} r_j(k) = r_j\mathbf{1}. \tag{6.52}$$

Thus, the definition of a probability matrix as a matrix with the sum of the entries in each row equal to 1 can be

[4]Note that r_j is an $1 \times n$ vector and $\mathbf{1}$ is an $n \times 1$ vector, and therefore the expression $r_j\mathbf{1}$ from (6.52) is consistent.

written as

$$\begin{aligned} \sum_{k=1}^{n} r_j(k) &= 1, \ \forall\, j = 1:n \\ \iff \quad r_j\mathbf{1} &= 1, \ \forall\, j = 1:n \\ \iff \quad (r_j\mathbf{1})_{j=1:n} &= \mathbf{1} \\ \iff \quad M\mathbf{1} &= \mathbf{1}, \end{aligned}$$

since $M\mathbf{1} = (r_j\mathbf{1})_{j=1:n}$ if $M = \mathrm{row}\,(r_j)_{j=1:n}$ is the row form of M.

In other words, we established that (6.51) is an equivalent condition for M to be a probability matrix.

Let A and B be probability matrices. Then all the entries of A and B are nonnegative, and therefore all the entries of AB are also nonnegative. From (6.51), it follows that

$$A\mathbf{1} = \mathbf{1} \quad \text{and} \quad B\mathbf{1} = \mathbf{1},$$

and therefore

$$(AB)\mathbf{1} = A(B\mathbf{1}) = A\mathbf{1} = \mathbf{1}.$$

Then, from (6.51), we conclude that AB is a probability matrix. □

Question 10. Find all the values of ρ such that

$$\begin{pmatrix} 1 & 0.6 & -0.3 \\ 0.6 & 1 & \rho \\ -0.3 & \rho & 1 \end{pmatrix}$$

is a correlation matrix.

Answer: Recall that a solution to this question based on was Sylvester's criterion was included in Chapter 2. We give two more solutions to this question here, one using the Cholesky decomposition, and another one based on the definition of symmetric positive semidefinite matrices.

A symmetric matrix with diagonal entries equal to 1 is a correlation matrix if and only if the matrix is symmetric positive semidefinite. Thus, we need to find all the values of ρ such that the matrix

$$\Omega = \begin{pmatrix} 1 & 0.6 & -0.3 \\ 0.6 & 1 & \rho \\ -0.3 & \rho & 1 \end{pmatrix} \tag{6.53}$$

is symmetric positive semidefinite.

Solution 1: To identify the values of ρ such that the matrix Ω is symmetric positive semidefinite, we apply the first step of the Cholesky algorithm to Ω, and obtain the following 2×2 matrix:

$$\begin{aligned} & \begin{pmatrix} 1 & \rho \\ \rho & 1 \end{pmatrix} - \begin{pmatrix} 0.6 \\ -0.3 \end{pmatrix} (0.6 \;\; -0.3) \\ = & \begin{pmatrix} 1 & \rho \\ \rho & 1 \end{pmatrix} - \begin{pmatrix} 0.36 & -0.18 \\ -0.18 & 0.09 \end{pmatrix} \\ = & \begin{pmatrix} 0.64 & \rho + 0.18 \\ \rho + 0.18 & 0.91 \end{pmatrix}. \end{aligned}$$

Thus, the matrix Ω is symmetric positive semidefinite if and only if the matrix

$$M = \begin{pmatrix} 0.64 & \rho + 0.18 \\ \rho + 0.18 & 0.91 \end{pmatrix}$$

is symmetric positive semidefinite. Since $M(1,1) = 0.64 > 0$, it follows that M is symmetric positive semidefinite if and only if $\det(M) \geq 0$, i.e., if and only if

$$\det(M) = 0.5824 - (\rho + 0.18)^2 \geq 0, \tag{6.54}$$

which is equivalent to

$$|\rho + 0.18| \leq \sqrt{0.5824} = 0.7632.$$

We conclude that Ω is a symmetric positive semidefinite matrix, and therefore a correlation matrix, if and only if

$$-0.7632 \leq \rho + 0.18 \leq 0.7632,$$

which can be written as

$$-0.9432 \leq \rho \leq 0.5832. \tag{6.55}$$

Note that condition (6.55) is the same as condition (2.9) obtained when solving the same question using Sylvester's Criterion; see Chapter 2.

Solution 2: By definition, the matrix Ω is symmetric positive semidefinite if and only if $x^t \Omega x \geq 0$ for all $x = (x_i)_{i=1:3} \in \mathbb{R}^3$. Note that

$$\begin{aligned} & x^t \Omega x \\ = \;& (x_1 \; x_2 \; x_3) \begin{pmatrix} 1 & 0.6 & -0.3 \\ 0.6 & 1 & \rho \\ -0.3 & \rho & 1 \end{pmatrix} \begin{pmatrix} x_1 \\ x_2 \\ x_3 \end{pmatrix} \\ = \;& x_1^2 + x_2^2 + x_3^2 + 1.2 x_1 x_2 - 0.6 x_1 x_3 + 2\rho x_2 x_3. \end{aligned}$$

By completing the square, we obtain that

$$\begin{aligned} & x^t \Omega x \\ = \;& x_1^2 + 2x_1(0.6x_2 - 0.3x_3) + x_2^2 + x_3^2 + 2\rho x_2 x_3 \\ = \;& (x_1 + 0.6x_2 - 0.3x_3)^2 \\ & -(0.6x_2 - 0.3x_3)^2 + x_2^2 + x_3^2 + 2\rho x_2 x_3 \\ = \;& (x_1 + 0.6x_2 - 0.3x_3)^2 \\ & +0.64x_2^2 + 2x_2x_3(\rho + 0.18) + 0.91x_3^2. \end{aligned}$$

By completing the square once again, we find that

$$\begin{aligned} & 0.64x_2^2 + 2x_2x_3(\rho + 0.18) + 0.91x_3^2 \\ = & \left(0.8x_2 + x_3\frac{\rho + 0.18}{0.8}\right)^2 \\ & -x_3^2\frac{(\rho + 0.18)^2}{0.64} + 0.91x_3^2 \\ = & \left(0.8x_2 + x_3\frac{\rho + 0.18}{0.8}\right)^2 \\ & +\frac{x_3^2}{0.64}\left(0.5824 - (\rho + 0.18)^2\right), \end{aligned}$$

and therefore

$$\begin{aligned} & x^t\Omega x \\ = & (x_1 + 0.6x_2 - 0.3x_3)^2 \\ & + \left(0.8x_2 + x_3\frac{\rho + 0.18}{0.8}\right)^2 \\ & +\frac{x_3^2}{0.64}\left(0.5824 - (\rho + 0.18)^2\right). \end{aligned}$$

Thus, $x^t\Omega x \geq 0$ for all $x = (x_i)_{i=1:3} \in \mathbb{R}^3$ if and only if

$$\begin{aligned} & (x_1 + 0.6x_2 - 0.3x_3)^2 \\ & + \left(0.8x_2 + x_3\frac{\rho + 0.18}{0.8}\right)^2 \\ & +\frac{x_3^2}{0.64}\left(0.5824 - (\rho + 0.18)^2\right) \\ \geq & 0, \ \forall\, x_1, x_2, x_3 \in \mathbb{R}. \end{aligned}$$

The last inequality holds if and only if

$$0.5824 - (\rho + 0.18)^2 \ \geq \ 0, \tag{6.56}$$

which is the same as (6.54).

We conclude that Ω is a correlation matrix if and only if

$$-0.9432 \leq \rho \leq 0.5832. \quad \square$$

6.3 Financial instruments: options, bonds, swaps, forwards, futures.

Question 1. The prices of three put options with strikes 40, 50, and 70, but otherwise identical, are \$10, \$20, and \$30, respectively. Is there an arbitrage opportunity present? If yes, how can you make a riskless profit?

Answer: If an arbitrage exists, it will be due to the fact that the convexity of put option values with respect to the strike price is violated.

In the plane (K, y), the line passing through the points $(K = 40, P(40) = 10)$ and $(K = 70, P(70) = 30)$ is given by

$$y = \frac{70 - K}{30} \cdot 10 + \frac{K - 40}{30} \cdot 30. \qquad (6.57)$$

The point on this line corresponding to strike 50 is obtained by substituting $K = 50$ in (6.57), and has y–coordinate equal to

$$\frac{2}{3} \cdot 10 + \frac{1}{3} \cdot 30 = \frac{50}{3}.$$

Since $P(K)$ is a strictly convex function of K, a no–arbitrage value of the put option with strike 50 should be below the line passing through the price points of the options with strikes 40 and 70. However, $P(50) = 20 > \frac{50}{3}$. Thus, the put option with strike 50 is overpriced, and an arbitrage exists.

Using a "buy low, sell high" strategy, we can take advantage of this arbitrage opportunity as follows: buy 2 put options with strike 40, buy 1 put option with strike 70, and sell 3 put options with strike 50. There is a \$10 positive cash flow when setting up this portfolio, since

$$3 \cdot \$20 \; - \; 2 \cdot \$10 \; - \; \$30 \; = \; \$10.$$

The value $V(T)$ of the portfolio at the maturity T of the options is

$$\begin{aligned} V(T) &= 2\max(40 - S(T), 0) \\ &\quad + \max(70 - S(T), 0) \\ &\quad - 3\max(50 - S(T), 0). \end{aligned}$$

Note that $V(T)$ is nonnegative for any value $S(T)$ of the underlying asset at T:

If $70 \leq S(T)$, then all options expire out of the money and

$$V(T) = 0.$$

If $50 \leq S(T) < 70$, then

$$V(T) = 70 - S(T) \geq 0.$$

If $40 \leq S(T) < 50$, then

$$\begin{aligned} V(T) &= (70 - S(T)) - 3(50 - S(T)) \\ &= 2S(T) - 80 \\ &\geq 0. \end{aligned}$$

If $S(T) < 40$, then

$$V(T) = 2S(T) - 80 + 2(40 - S(T)) = 0.$$

In other words, we set up a portfolio with positive cash flow at inception which does not lose money regardless of the value of the underlying asset at time T. The risk–free profit is equal to the future value at time T of the \$10 cash flow from setting up the portfolio. $\square$

Question 2. The price of a stock is \$50. In three months, it will either be \$47 or \$52, with 50% probability. How much would you pay for an at the money put? Assume for simplicity that the stock pays no dividends and that interest rates are zero.

Answer: Recall first that real world probabilities do not play any role in valuing an option in a (one period) binomial tree model. Thus, the 50% probability stated in the question is only meant to throw you off–course.

Solution 1: The value of the option is the discounted expected value of the payoff of the option in the risk–neutral probability measure. Since interest rates are zero, this can be written as

$$P(0) = p_{RN,up}P_{up} + p_{RN,down}P_{down}. \tag{6.58}$$

The up and down factors are $u = \frac{52}{50} = 1.04$ and $d = \frac{47}{50} = 0.94$, respectively. The risk–neutral probabilities of going up and down are

$$p_{RN,up} = \frac{1-d}{u-d} = 0.6; \quad p_{RN,down} = \frac{u-1}{u-d} = 0.4.$$

The ATM put pays \$3 if the stock price goes down to \$47, i.e., $P_{down} = 3$, and expires worthless if the stock price goes up to \$52, i.e., $P_{up} = 0$. From (6.58), we find that the value of the ATM put is

$$P(0) = 0.6 \cdot 0 + 0.4 \cdot 3 = 1.2. \tag{6.59}$$

In other words, you should pay at most \$1.20 for an at the money put.

Solution 2: An insightful solution can be given by setting up a hedged portfolio. The Delta of the put option is

$$\Delta_P = \frac{P_{up} - P_{down}}{S_{up} - S_{down}} = \frac{0-3}{52-47} = -0.6.$$

A portfolio which is long one ATM put and short Δ_P shares will be long the put and long 0.6 shares, and will have the same value at maturity regardless of whether the stock price goes down to \$47 or up to \$52:

$$\begin{aligned}
\Pi(T) &= P(T) + 0.6S(T) \\
&= \begin{cases} 0 + 0.6 \cdot 52 = 31.20, & \text{if } S(T) = 52; \\ 3 + 0.6 \cdot 47 = 31.20, & \text{if } S(T) = 47. \end{cases}
\end{aligned}$$

For no–arbitrage, the value of the portfolio at inception must be the discounted value of its payoff. Since the interest rates are zero, we obtain that $\Pi(0) = P(0) + 0.6S(0) = 31.20$, and therefore

$$P(0) = 31.20 - 0.6 \cdot 50 = 1.20,$$

which is the same value of the put, \$1.20, obtained above; see (6.59). □

Question 3. A stock worth \$50 today will be worth either \$60 or \$40 in three months, with equal probability. The value of a three months at the money put on this stock is \$4. Does the value of the three months ATM put increase or decrease, and by how much, if the probability of the stock going up to \$60 were 75% and the probability of the stock going down to \$40 were 25%?

Answer: In a one period binomial tree model, the actual probabilities of the asset going up or down do not play any role in the valuation of a plain vanilla option. Thus, the value of the three months at the money put would be the same, \$4, even if the probability of the stock going up to \$60 were 75%. □

Question 4. What is risk–neutral pricing?

Answer: Risk–neutral pricing, or valuation, refers to valuing derivative securities as discounted expected values of their payoffs at maturity, under the assumption that the underlying asset has lognormal distribution with a drift equal to the risk-free rate.

More precisely, if the price of the underlying asset has a lognormal distribution with volatility σ and pays dividends continuously at the rate q, then the value of a derivative security on this asset with payoff $V(T)$ at maturity T given by risk–neutral valuation is

$$V(0) = e^{-rT} E_{RN}[V(S(T))],$$

where r is the risk–free rate assumed to be constant, and the expected value is computed with respect to the lognormal random variable $S(T)$ given by

$$S(T) = S(0)\, e^{\left(r-q-\frac{\sigma^2}{2}\right)T + \sigma\sqrt{T}Z}.$$

Risk–neutral valuation can be used for derivative securities which can be perfectly hedged dynamically using cash and the underlying asset. Plain vanilla European options, as well as European options with other payoffs at maturity (such as asset–or–nothing and cash–or–nothing options) can be priced using risk–neutrality. Risk–neutral valuation cannot be used for path dependent options such as American options, barrier options, and Asian options. □

Question 5. Describe briefly how you arrive at the Black–Scholes formula.

Answer: The Black–Scholes formulas give the values of plain vanilla European put and call options on an underlying asset with lognormal distribution. Several methods for deriving the Black-Scholes formulas are:

• Risk neutral pricing: the expected value of the payoff of the option at maturity computed under the assumption that the price of the underlying asset has a lognormal distribution with drift equal to the risk free rate gives the Black–Scholes value of the option.

• Black–Scholes PDE solution: the Black–Scholes value of the option satisfies the Black–Scholes PDE with boundary conditions given by the payoff of the option at maturity. The Black–Scholes PDE is transformed into the heat PDE using a lognormal change of variables, and the closed form solution of the heat PDE is then used to derive the closed form solution of the Black–Scholes PDE, which is the Black–Scholes value of the option.

• Binomial tree model pricing: the evolution of the underlying asset is modeled using a binomial tree calibrated to converge in the limit to a lognormal distribution with drift equal to the risk free rate. For every tree, an approximate option value is obtained from the binomial tree model. The limit of these binomial tree option values as the number of time intervals in the tree goes to infinity is the Black–Scholes value of the option.

Note that twelve different ways to derive the Black–Scholes formula can be found in Wilmott [4]. □

Question 6. How much should a three months at the money put on an asset with spot price \$40 and volatility 30% be worth? Assume, for simplicity, that interest rates are zero and that the asset does not pay dividends.

Answer: The following approximation for the value of an at the money put option on a non dividend paying underlying asset and assuming zero risk–free interest rates is easy to estimate and very accurate if the total variance is small (e.g., if $\sigma^2 T \leq 0.25$):

$$P_{ATM} \approx 0.4\sigma S_0 \sqrt{T}; \tag{6.60}$$

see Stefanica [3] for a derivation of formula (6.60).

For $S_0 = 40$, $\sigma = 0.3$, and $T = \frac{1}{4}$, we obtain that the value of the at the money put is approximately 2.40.

For comparison purposes, note that the value of the put option computed using the Black–Scholes formula would be 2.3914; the approximate formula (6.60) is very accurate in this case. □

Question 7. If the price of a stock doubles in one day, by how much will the value of a call option on this stock change?

Answer: The value of a deep in the money call on a non dividend paying asset can be approximated, e.g., by using the Put-Call parity, as $C \approx S - Ke^{-rT}$, where K and T are the strike and the maturity of the option, and r is the constant risk free rate. Thus, if the spot price S doubles, the call option will be even deeper in the money, and therefore its value will be approximately $2S - Ke^{-rT}$. In other words, the value of deep in the money calls roughly doubles if the spot price doubles.

If the option is around at the money, the percentage change generated by the doubling of the stock price is about one order of magnitude larger since the option will become deep in the money.

If the option is deep out of the money, then it trades for fraction of cents. The doubling of the spot price would result in changing of the price of the option by several orders of magnitude.

As a numerical example, consider a six months call option with strike 20 on a non dividend paying underlying asset with volatility 25%. Assume that the risk free rate is constant at 5%. The Black–Scholes values of the call option corresponding to several spot prices of the underlying asset can be found below:

Spot Price	Option Price
10	0.000045
20	1.65
40	20.49
80	60.49
400	380.49
800	780.49

If the call option is deep out of the money and the spot price doubles from \$10 to \$20, the value of the call increases from \$0.000045 to \$1.65, i.e., by more than four orders of magnitude.

If the call option is at the money and the spot price doubles from \$20 to \$40, the value of the call increases from \$1.65 to \$20.49, i.e., more than tenfold.

If the call option is deep in the money, and the spot price doubles from \$40 to \$80, the value of the call increases from \$20.49 to \$60.49, i.e., by a factor of 2.95; if the call option is even deeper in the money[5] and the spot price doubles from \$400 to \$800, the value of the call increases from \$380.49 to \$780.49, i.e., the call approximately doubles in value.

Moreover, if the call option is deep in the money, its value is very close to $S - Ke^{-rT}$, i.e., the value of the spot price of the underlying asset minus the present value of the strike. For all the spot prices greater than \$40, the estimate $C \approx S - Ke^{-rT}$ is very accurate. Thus, if the call option is at the money and the spot price doubles, the value of the call option increases by the same amount as the increase in the spot price. For example, if the spot price doubles from \$40 to \$80, the value of the call increases by \$40, from \$20.49 to \$60.49, which is exactly the increase in the spot price. □

Question 8. What are the smallest and largest values that Delta can take?

Answer: Assume, for simplicity, that the underlying asset does not pay dividends.

The Delta (Δ) of a long position in a plain vanilla call option is between 0 and 1 (and therefore the Delta of a short plain vanilla call position is between -1 and 0). The Delta of a long call position increases with the spot price of the underlying asset, and goes from 0, when the asset is worthless (i.e., when the call option is deep out of

[5] Of course, call options with strike ten times smaller than the spot price of the underlying asset never occur in practice; this part of the example is for illustration purposes.

the money), to 1, when the spot price of the asset is very large (i.e., when the call option is deep in the money).

The Delta of a long position in a plain vanilla put option is between -1 and 0; the Delta of a short position in a plain vanilla put option is between 0 and 1. The Delta of a long put position also increases with the spot price of the underlying asset, and goes from -1, when the asset is worthless (i.e., when the put option is deep in the money), to 0, when the spot price of the asset is very large (i.e., when the put option is deep out of the money).

Note that, from the Put-Call parity, i.e., $C - P = S - Ke^{-rT}$, it follows that $\Delta(P) = \Delta(C) - 1$, which is consistent with the bounds above. □

Question 9. What is the Delta of an at-the-money call? What is the Delta of an at-the-money put?

Answer: The Delta of an at-the-money call is approximately 0.5; the Delta of an at-the-money put is approximately -0.5.

Assume, for simplicity, a Black-Scholes framework with zero risk-free rates and an underlying asset paying no dividends. Then, $\Delta(C_{BS}) = N(d_1)$, where $N(x)$ denotes the cumulative distribution of the standard normal variable and

$$d_1 = \frac{\ln\left(\frac{S}{K}\right) + \left(r - q + \frac{\sigma^2}{2}\right)T}{\sigma\sqrt{T}} = \frac{\sigma\sqrt{T}}{2} \qquad (6.61)$$

for $K = S$ and for $r = q = 0$.

The linear Taylor approximation of $N(x)$ around 0 is

$$N(x) \approx 0.5 + \frac{x}{\sqrt{2\pi}} \approx 0.5 + 0.4x, \qquad (6.62)$$

since $\frac{1}{\sqrt{2\pi}} = 0.3989 \approx 0.4$. Thus,

$$\Delta(C_{BS}) = N(d_1) = N\left(\frac{\sigma\sqrt{T}}{2}\right) \approx 0.5 + 0.2\sigma\sqrt{T}.$$

This is roughly estimated as $\Delta(C_{BS}) \approx 0.5$, since $0.2\sigma\sqrt{T}$ is small for most options (e.g., $0.2\sigma\sqrt{T} \leq 0.1$ for volatility less than 50% and for maturity less than one year).

For put options, $\Delta(P_{BS}) = -N(-d_1)$, where $d_1 = \frac{\sigma\sqrt{T}}{2}$; cf. (6.61). From (6.62), it follows that

$$\Delta(P_{BS}) = -N\left(-\frac{\sigma\sqrt{T}}{2}\right) \approx -0.5 + 0.2\sigma\sqrt{T},$$

which, using the same rationale outlined above, is often stated as $\Delta(P_{BS}) \approx -0.5$. □

Question 10. What is the Put–Call parity? How do you prove it?

Answer: The Put–Call parity is a model independent no–arbitrage relationship between the prices of European call and put options with the same strike and maturity.

In a nutshell, the Put–Call parity states that being long a call option and short a put option on the same underlying asset, and with the same strike and maturity, is the same as being long a forward contract on the asset, with the same maturity as the maturity of the options, and with delivery price equal to the strike of the options. Equivalently, being long a call and short a put is the same as being long one unit of the underlying asset (for non dividend paying assets) and short the present value of the strike of the options.

More precisely, if $C(t)$ and $P(t)$ are the values at time t of a European call and put option with maturity T and strike K, on the same non dividend paying asset with spot price $S(t)$, then the Put–Call parity states that

$$C(t) - P(t) = S(t) - Ke^{-r(T-t)}, \tag{6.63}$$

where r denotes the risk–free rate, assumed to be constant.

If the underlying asset pays dividends continuously at the rate q, the Put–Call parity has the form

$$C(t) - P(t) = S(t)e^{-q(T-t)} - Ke^{-r(T-t)}. \tag{6.64}$$

For simplicity, we restrict our attention to non dividend paying underlying assets and prove formula (6.63).

Consider a portfolio made of the following assets:

- long 1 put option;
- long 1 unit of the underlying asset;
- short 1 call option.

The value of the portfolio at time t is

$$V_{portfolio}(t) = P(t) + S(t) - C(t). \tag{6.65}$$

The values of the call and put option at maturity are

$$\begin{aligned} C(T) &= \max(S(T) - K, 0); \\ P(T) &= \max(K - S(T), 0). \end{aligned}$$

Then, regardless of whether $S(T) < K$ or $S(T) \geq K$, the value of the portfolio at time T will be equal to K:

If $S(T) < K$, then $P(T) = K - S(T)$ and $C(T) = 0$, and therefore

$$\begin{aligned} V_{portfolio}(T) &= P(T) + S(T) - C(T) \\ &= (K - S(T)) + S(T) - 0 \\ &= K. \end{aligned} \tag{6.66}$$

If $S(T) \geq K$, then $P(T) = 0$ and $C(T) = S(T) - K$, and therefore

$$\begin{aligned} V_{portfolio}(T) &= P(T) + S(T) - C(T) \\ &= 0 + S(T) - (S(T) - K) \\ &= K. \end{aligned} \tag{6.67}$$

From (6.66) and (6.67), we find that

$$V_{portfolio}(T) = P(T) + S(T) - C(T) = K, \tag{6.68}$$

regardless of the value $S(T)$ of the underlying asset at the maturity of the option.

Then, for no–arbitrage, the value of the portfolio at time t must be equal to the present value of K at time t, i.e.,

$$V_{portfolio}(t) = Ke^{-r(T-t)}. \qquad (6.69)$$

From (6.65) and (6.69), we obtain that

$$P(t) + S(t) - C(t) = Ke^{-r(T-t)}.$$

This can be written as

$$C(t) - P(t) = S(t) - Ke^{-r(T-t)},$$

which is the Put–Call parity formula (6.63).

The Put–Call parity formula (6.64) corresponding to an underlying asset paying dividends continuously at the rate q can be obtained similarly using a portfolio with a long put position, a short call position, and a long position in $e^{-q(T-t)}$ units of the underlying asset. All dividends payed by the long asset position between time t and time T are used to purchase additional fractions of the asset. Doing so continuously results in an asset position at time T equal to long one unit of the asset. Thus, the value of the portfolio at time T will be equal to K regardless of the value of $S(T)$, and the Put–Call parity formula (6.64) for underlying assets paying dividends continuously can be obtained as before. □

Question 11. Show that the time value of a European call option is highest at the money.

Answer: The time value of a call option is the difference between the value $C(S)$ of the option and the intrinsic value $\max(S-K, 0)$ of the call option. In other words, the time value of the option is

$$C(S) - \max(S-K, 0).$$

We want to establish that the time value of the option is highest at the money, i.e., when $S = K$. To do so, we show that the function

$$f(S) = C(S) - \max(S - K, 0)$$

attains its maximum for $S = K$.

Note that[6]

$$f(S) = \begin{cases} C(S), & \text{if } S \leq K; \\ C(S) - S + K, & \text{if } S > K. \end{cases}$$

For $S \leq K$, the function $f(S)$ is the value of a call with strike K, and therefore is increasing.

For $S > K$, we find that

$$f'(S) = \Delta(C) - 1 < 0,$$

since the Delta of a call option is always less than 1,[7] and therefore the function $f(S)$ is decreasing.

We conclude that $f(S)$ has a global maximum point at $S = K$, which is what we wanted to show.

Note that a similar reasoning shows that the time value of a put option, given by

$$P(S) - \max(K - S, 0),$$

---

[6]The function $f(S)$ is a continuous function, but it is not differentiable at $S = K$.

[7]The Delta of a long call position is an increasing function going from 0 when the spot price of the underlying asset is 0 and the call option is deep out of the money, to 1 when the spot price of the underlying asset goes to ∞ and the call option is deep in the money. For example, in the Black–Scholes model,

$$\Delta(C_{BS}) = e^{-qT} N(d_1) < N(d_1) < 1.$$

is largest at the money, i.e., for $S = K$. If $g(S) = P(S) - \max(K - S, 0)$, then

$$g(S) = \begin{cases} P(S) - K + S, & \text{if } S \leq K; \\ P(S), & \text{if } S > K. \end{cases}$$

For $S \leq K$,

$$g'(S) = \Delta(P) + 1 > 0,$$

and therefore the function $g(S)$ is increasing, since the Delta of a put option is between -1 (when the put is deep in the money) and 0 (when the put is deep out of the money), and therefore is always greater than -1.

For $S > K$, the function $g(S)$ is the value of a put with strike K, and therefore is decreasing.

Thus, $g(S)$ has a global maximum point at $S = K$, and therefore the time value of the put option is largest at the money. □

Question 12. What is implied volatility? What is a volatility smile? How about a volatility skew?

Answer: By definition, implied volatility is the unique value of the volatility parameter σ from the lognormal model for the evolution of the price of an underlying that makes the Black–Scholes value of an option equal to the market price of the option. Implied volatility exists and is unique[8] for any arbitrage–free market value of the option.

On the same asset, prices of options with multiple strikes and maturities are quoted, and implied volatilities can be computed for each of these options. If the price of the asset had a lognormal distribution as assumed in the

[8]The uniqueness of the implied volatility comes from the fact that the Black–Scholes value of the option is a strictly increasing function of the volatility parameter (or, equivalently, the vega of the option is strictly positive).

Black–Scholes model, then the resulting plots of implied volatility vs strike for the same maturity should be flat. In practice, they are not flat, and are often shaped as "smiles" or "skews".

An implied volatility smile occurs when the implied volatilities of deep in the money options and of deep out of the money options are higher than the implied volatilities of options close to at the money. Volatility smiles are typical for currency options.

An implied volatility skew occurs when the implied volatilities of options with large strikes are lower than the implied volatilities of at the money options (reverse skew), or when the implied volatilities of options with small strikes are lower than the implied volatilities of at the money options (forward skew). Reverse skews are typical for long dated equity and index options. Forward skews are typical for commodities options. □

Question 13. What is the Gamma of an option? Why is it preferable to have small Gamma? Why is the Gamma of plain vanilla options positive?

Answer: The Gamma (Γ) of an option measures the sensitivity of the Delta of the option with respect to the price of the underlying asset, i.e.,

$$\Gamma(V) = \frac{\partial \Delta}{\partial S} = \frac{\partial^2 V}{\partial S^2},$$

where V denotes the value of the option.

It is often important to immunize a portfolio with respect to changes in the price of the underlying asset, i.e., to make the portfolio Delta–neutral. A portfolio with small Gamma would need to be rebalanced less often in order to be kept Delta–neutral, since the change in the Delta of the portfolio is proportional to Gamma for small changes in the value of the underlying asset. Thus, a

Delta–neutral and Gamma–neutral portfolio is well hedged against small changes in the price of the underlying asset (although not against jumps in the price of the underlying asset).

The Delta of plain vanilla options (calls or puts) increases as the spot price of the underlying asset increases. Thus, Gamma is positive, since Gamma is the rate of change of Delta. Moreover, Gamma is asymptotically going to 0 for deep out of the money options and for deep in the money options, and the highest value of Gamma corresponds to options with strike close to the spot price (at the money options). □

Question 14. When are a European call and a European put worth the same? (The options are written on the same asset and have the same strike and maturity.) What is the intuition behind this result?

Answer: Recall from the Put–Call parity that

$$C - P = Se^{-qT} - Ke^{-rT}, \qquad (6.70)$$

where C and P are the values of a call and put option, respectively, with strike K and maturity T on an underlying asset with spot price S and paying dividends continuously at rate q. If $C = P$, it follows from (6.70) that $K = Se^{(r-q)T}$. Since the forward price of the underlying asset is $F = Se^{(r-q)T}$, we conclude that a call and a put are worth the same if their strike is equal to the forward price of the asset; these options are called at–the–money–forward options.

Note that this result is independent of any assumption on the evolution of the price of the underlying asset, since Put–Call parity is model independent.

The fact that an at–the–money–forward call and an at–the–money–forward put are worth the same may seem counterintuitive at first glance: call options have unlimited upside since their payoff at maturity,

$$C(T) = \max(S(T) - K, 0),$$

can be infinitely large, while put options have limited upside since their payoff at maturity,

$$P(T) = \max(K - S(T), 0),$$

is bounded by the strike price, i.e., $P(T) \leq K$.

However, the value of an option is equal to the risk–neutral expected value of the option payoff at maturity. In every model for the evolution of the price of the underlying asset (including in the geometric Brownian motion model underlying the Black–Scholes framework), the probability density of the underlying asset at maturity decreases exponentially for large values of the spot price. This renders the expected value of large payoffs negligible, and makes it possible for at–the–money–forward put options to be worth the same as at–the–money–forward call options. □

Question 15. What is the two year volatility of an asset with 30% six months volatility?

Answer: Asset volatility scales with the square root of time: if σ denotes the annualized volatility of an asset, the volatility $\sigma(t)$ of the asset over a time horizon t is given by $\sigma(t) = \sigma\sqrt{t}$. Then,

$$\frac{\sigma(t_2)}{\sigma(t_1)} = \sqrt{\frac{t_2}{t_1}}.$$

For $t_2 = 2$, $t_1 = 0.5$, and $\sigma(t_1) = \sigma(0.5) = 0.3$, we find that $\sigma(t_2) = \sigma(2) = 0.6$, i.e., the two year volatility of the asset is 60%. □

Question 16. How do you value an interest rate swap?

Answer: For valuation purposes only, add payments at maturity equal to the notional of the swap both for the fixed leg and for the floating leg of the swap. Then, the value of the swap for the party receiving fixed payments is $V_{swap} = V_{fix} - V_{float}$, where V_{fix} is the value of a coupon bond with coupon rate equal to the fixed rate of the swap, and V_{float} is the value of an instrument making the floating payments of the swap, plus a payment equal to the notional at maturity.

The value V_{fix} of the fixed rate coupon bond is the sum of the present values of its future payments discounted using risk–free zero rates (most frequently, LIBOR rates).

To compute V_{float}, note that, right after a payment is made, the remaining payments of the floating leg (including the notional at maturity) are equivalent to rolling over the notional at the prevailing zero rates until maturity. Thus, the value of all the floating payments on a payment date for the swap is equal to the notional. We conclude that V_{float} is the present value of the next floating payment (which was determined at the prior swap payment date) plus the notional.

To illustrate swap valuation with an example, consider a 19–month semiannual swap on a \$10 million notional with 3% fixed rate and paying semiannually compounded LIBOR. The next floating payment that will be made in one month is \$125,000 (and was determined five months ago, at the previous cash flow date).

The cash flow dates of the swap are 1 month, 7 months, 13 months, and 19 months. The value of the swap for the party receiving fixed payments is $V_{swap} = V_{fix} - V_{float}$. The value of the 3% semiannual coupon bond corresponding to the fixed leg of the swap is

$$\begin{aligned} V_{fix} &= 150,000 \cdot \mathrm{disc}\left(\frac{1}{12}\right) \\ &\quad + 150,000 \cdot \mathrm{disc}\left(\frac{7}{12}\right) \\ &\quad + 150,000 \cdot \mathrm{disc}\left(\frac{13}{12}\right) \\ &\quad + 10,150,000 \cdot \mathrm{disc}\left(\frac{19}{12}\right), \end{aligned} \tag{6.71}$$

where $\mathrm{disc}(t)$ denotes the discount factor corresponding to time t. For example, if the discount factors are given in terms of the semiannually compounded LIBOR rate $\mathrm{LIBOR}(t)$, then

$$\mathrm{disc}(t) = \left(1 + \frac{\mathrm{LIBOR}(t)}{2}\right)^{-2t}.$$

On the next cash flow date, i.e., in one month, the floating leg of the swap is equal to the value of the next floating payment, \$125,000, plus the \$10 million notional, i.e., \$10,125,000. Thus, the value of the floating leg of the swap today is the present value of \$10,125,000 in one month, i.e.,

$$V_{float} = 10,125,000 \cdot \mathrm{disc}\left(\frac{1}{12}\right). \tag{6.72}$$

The value of the swap for the party receiving fixed payments is $V_{swap} = V_{fix} - V_{float}$, where V_{fix} and V_{float} are given by (6.71) and (6.72), respectively. □

Question 17. By how much will the price of a ten year zero coupon bond change if the yield increases by ten basis points?

Answer: The first order approximation of the change ΔB in the bond price in terms of the change Δy in the bond yield is

$$\Delta B \approx \frac{\partial B}{\partial y}\Delta y = -DB\Delta y, \qquad (6.73)$$

where $D = -\frac{1}{B}\frac{\partial B}{\partial y}$ is the duration of the bond. Thus,

$$\frac{\Delta B}{B} \approx -D\Delta y. \qquad (6.74)$$

Note that $D = 10$, since the duration of a zero coupon bond is equal to the maturity of the bond. Moreover, the change in the bond yield is $\Delta y = 0.001$, since $1\% = 100$ basis points (bps), and therefore 10 bps $= 0.1\% = 0.001$.

From (6.74), we find that the percentage change in the value of the bond can be estimated as follows:

$$\frac{\Delta B}{B} \approx -10 \times 0.001 = -0.01.$$

In other words, the price of the ten year zero-coupon bond decreases by 1% if the yield increases by ten basis points. □

Question 18. A five year bond with 3.5 years duration is worth 102. What is the value of the bond if the yield decreases by fifty basis points?

Answer: The value of the bond will increase, since the yield of the bond decreases. More precisely, recall from (6.73) that

$$\Delta B \approx -DB\Delta y. \qquad (6.75)$$

For $B = 102$, $D = 3.5$ and $\Delta y = -0.005$ ($1\% = 100$ bps, and therefore 50 bps $= 0.5\% = 0.005$), we find from (6.75) that $\Delta B \approx 1.785$. Thus, the new value of the bond is $B + \Delta B \approx 103.785$. □

Question 19. What is a forward contract? What is the forward price?

Answer: A forward contract is an agreement between two parties in which one party (the long position) agrees to buy a specified quantity of the underlying asset from the other party (the short position) at a given time in the future and for a price, called the forward price, that is agreed upon at the inception of the forward contract. The forward price is chosen such that the forward contract has value 0 at inception.[9]

If the underlying asset has spot price S_0 and pays dividends continuously at rate q, the forward price for a forward contract maturing at T is

$$F = S_0 e^{(r-q)T},$$

where r is the risk–free rate between 0 and T. □

Question 20. What is the forward price for treasury futures contracts? What is the forward price for commodities futures contracts?

Answer: A short position in a forward contract (i.e., selling a forward contract) is perfectly hedged by buying one unit of the underlying asset and holding that position until the maturity of the forward contract. The forward price is the future value at the maturity of the forward contract of the cost of buying one unit of the underlying asset.

For a treasury futures contract, buying the underlying treasury bond generates a positive cash flow from receiving all the bond coupon payments until the maturity of the forward contract. If S_0 is the spot price of the underlying treasury bond and if C is the present value of all the coupon payments received during the life of the futures contract, then the forward price F of a treasury futures contract with maturity T is the future value at time T of $S_0 - C$, i.e., $F = (S_0 - C)e^{rT}$.

[9] Note that the forward price is not the price of the forward.

For a commodities futures contract, buying the underlying commodity incurs storage costs; for example, for a gold futures contract, buying the underlying gold would require storing the gold safely. If S_0 is the spot price of the underlying commodity and if C is the present value of all the storage costs, then the forward price F of a commodities futures contract with maturity T is the future value at time T of $S_0 + C$, i.e., $F = (S_0 + C)e^{rT}$, where r denotes the risk–free rate between 0 and T. □

Question 21. What is a Eurodollar futures contract?

Answer: A Eurodollar is a dollar deposit in a bank outside the United States. The Eurodollar rate is the interest rate earned on Eurodollars deposited by one bank with another bank (and is very close to LIBOR for short term maturities).

A Eurodollar futures contract is a futures contract on a Eurodollar rate, and deliveries are for up to ten years in the future.

For example, a three-month Eurodollar futures contract is a futures contract on the three month (90-day) Eurodollar rate. The start date is the third Wednesday of the delivery month (March, June,
September, December). □

Question 22. What are the most important differences between forward contracts and futures contracts?

Answer: The main differences between the ways forward and futures contracts are structured, settled, and traded are:

• Futures contracts trade on exchanges and have standard features, while forward contracts are over-the-counter instruments.

• Futures are marked to market and settled in a margin account on a daily basis, while forward contracts are not settled before maturity.

• Futures carry almost no credit and counterparty risk, since they are settled daily, while entering into a forward contract carries some credit risk.

• Futures have a range of delivery dates, while forward contracts have a specified delivery date.

• Futures contracts require the delivery of the underlying asset for the futures price, while forward contracts can be settled in cash at maturity, without the delivery of a physical asset. □

Question 23. What is the ten–day 99% VaR of a portfolio with a five–day 98% VaR of $10 million?

Answer: If we assume normally distributed short term portfolio returns, the VaR (Value at Risk) of a portfolio is proportional to the square root of the time horizon. More precisely, if the N day $C\%$ VaR of the portfolio is denoted by $\mathrm{VaR}(N, C)$, where N is the number of days in the time horizon and C is the confidence level, then

$$\mathrm{VaR}(N, C) \approx \sigma_V z_C \sqrt{\frac{N}{252}}\, V(0), \qquad (6.76)$$

where, σ_V is the (annualized) standard deviation of the rate of return of the portfolio, z_C is the z–score of the standard normal distribution corresponding to C, i.e., $P(Z \leq z_C) = C$, and $V(0)$ is the current value of the portfolio.

From (6.76), it follows that

$$\mathrm{VaR}(10 \text{ days}, 99\%) \approx \frac{z_{99}\sqrt{10}}{z_{98}\sqrt{5}}\, \mathrm{VaR}(5 \text{ days}, 98\%).$$

For $\mathrm{VaR}(5 \text{ days}, 98\%) = \$10{,}000{,}000$, and since $z_{99} \approx 2.326348$ and $z_{98} \approx 2.053749$, we obtain that

$$\mathrm{VaR}(10 \text{ days}, 99\%) \approx \$16{,}019{,}255. \quad \square$$

6.4 C++. Data structures.

Question 1. How do you declare an array?

Answer: An array can be declared either on stack, or on heap.

```
//created on stack, uninitialized
T identifier[size];

//created on stack, initialized
T identifier[] = initializer_list;

//created on heap, uninitialized
T* identifier = new T[size];
```

Example:

```
int foo[3];
int bar[] = {1,2,3};
int* baz = new int[3];
```

Question 2. How do you get the address of a variable?

Answer: Use the ampersand before the name of the variable, e.g.

```
T var;
T* ptr = &var
```

Example:

```
int foo = 1;
int* foo_ptr = &foo;
```

Question 3. How do you declare an array of pointers?

Answer: The same way as declaring an array, but making the type, `T`, a pointer:

```
T* identifier[size];
T* identifier[] = initializer_list;
T** identifier = new T*[size];
```

Example:

```
int a = 1; int b = 2; int c = 3;
int* foo[3];
int* bar[] = {&a, &b, &c};
int** baz = new int*[3];
```

Question 4. How do you declare a const pointer, a pointer to a const and a const pointer to a const?

Answer:

```
//pointer to a read only variable
const T* identifier;
T const* identifier;

//read only pointer to a variable
T *const identifier = rvalue;

//read only pointer to a read only variable
const T *const identifier = rvalue;
T const *const identifier = rvalue;
```

Example:

```
//read only variables
const int a = 2; const int b = 2;
int c  = 1;

//pointer to a read only b
int const* foo_two;
foo_two = &a; foo_two = &b;

//pointer to read only a
```

```
const int* foo;
foo = &a; foo = &b;

//read only pointer to c
//it needs to be initialized
int *const bar = &c;

//read only pointer to read only a
//it needs to be initialized
const int *const baz = &a;
```

Question 5. How do you declare a dynamic array?

Answer:

```
T* identifier = new T[size];
T* identifier = nullptr;
T* identifier;

delete[] identifier;
```

Example:

```
int *foo = new int[4];
int *bar = nullptr;
bar = new int[4];
```

Question 6. What is the general form for a function signature?

Answer:

```
return_type function_name(parameter_list);
```

Example:

```
int my_sum(int a, int b);
```

Question 7. How do you pass-by-reference?

Answer:

```
return_type function_name(T & identifier);
```

The identifier is now an alias for the argument.

Question 8. How do you pass a read only argument by reference?

Answer:

```
return_type function_name(const T & identifier);
```

Once you define a parameter as `const`, you will not be able to modify it in the function.

Question 9. What are the important differences between using a pointer and a reference?

Answer: Several differences between using a pointer and a reference are:

• A pointer can be re-assigned any number of times, while a reference cannot be reassigned after initialization.

• A pointer can point to NULL (`nullptr` in C++11), while a reference can never be referred to NULL.

• It is not possible to take the address of a reference as it is done with pointers.

• There is no reference arithmetic.

Question 10. How do you set a default value for a parameter?

Answer:

```
return_type function_name(T identifier = rvalue)
```

The parameters with default value must be placed at the end of the parameter list.

Question 11. How do you create a template function?

Answer:

```
template<class T>
return_type function_name(parameter_list);

template<typename T>
return_type function_name(parameter_list);
```

Note that the parameter type can be specified, when calling the function, explicitly or implicitly. Also note that there is no technical difference between using `class` or `typename` besides code readability (`typename` for primitive types and `class` for classes).

Example:

```
template<typename T>
T temp_sum(T a, T b) {return a+b;}

struct Processor{
    int a;
    int apply(int b) {return a+b;}
};

template<class T>
int temp_sum_2(int a, int b) {
    T processor;
    processor.a = a;
    return processor.apply(b);
}

int main(){
    //implicit, foo equals 3
    int foo = temp_sum(1,2);

    //explicit, bar equals 3
    int bar = temp_sum_2<Processor>(1,2);
}
```

Question 12. How do you declare a pointer to a function?

Answer:

```
return_type (*identifier) (list_parameter_types)
```

Example:

```
int my_sum(int a, int b) {return a+b;}
int main(){
    int(*p_func)(int,int);
    p_func = & my_sum;

    // foo equals 3
    int foo = p_func(1,2);
}
```

Question 13. How do you prevent the compiler from doing an implicit conversion with your class?

Answer: Use the keyword `explicit` to define the constructor:

```
explicit Classname(parameter_list)
```

Question 14. Describe all the uses of the keyword `static` in C++.

Answer: Inside a function, using the keyword `static` means that once the variable has been initialized it remains in memory up until the end of the program.

Inside a class definition, either for a variable or for a member function, using the keyword `static` means that the there is only one copy of them per class, and shared between instances.

As a global variable in a file of code, using the keyword `static` means that the variable is private within the scope of the file.

Question 15. Can a static member function be const?

Answer: When the `const` qualifier is applied to a non-static member function it implies that member function can not change the instance class when called (i.e. can not change any non `mutable` members from `*this`). Since static member function are defined at a class level, where there is no notion of `this` the `const` qualifier for member functions does not apply.

Question 16. C++ constructors support the initialization of member data via an initializer list. When is this preferable to initialization inside the body of the constructor?

Answer: The initialization list has to be used for const members, references and with members without default constructors, but for any type of members initialization through the initialization list is still preferable, since it is for efficient. Using the initialization list, the members are initialized calling directly their constructors. If the initialization is done in the body of the constructor for each member being initialized there is an instance of it created and then a copy assignment operation is called to assign that instance to its respective member.

Question 17. What is a copy constructor, and how can the default copy constructor cause problems when you have pointer data members?

Answer: A copy constructor allows you to create a new object as a copy of an existing instance. The default copy constructor creates the new object by copying the existing object, member by member, and thus when there are member pointer you end up with two objects pointing to the same object.

It is important to note that the copy constructor is called every time a function receives an object via the pass-by-value mechanism. This means that the copy constructor needs to be implemented using a pass by reference. Otherwise you will be recursively calling the copy constructor. You should always set the parameter for a copy constructor to be const.

```
ClassName( const ClassName& other );
ClassName( ClassName& other );
ClassName( volatile const ClassName& other);
ClassName( volatile ClassName& other );
```

Question 18. What is the output of the following code:

```
#include <iostream>
using namespace std;

class A
{
public:
    int * ptr;
    ~A()
    {
        delete(ptr);
    }
};

void foo(A object_input)
{
    ;
}

int main()
{
    A aa;
    aa.ptr = new int(2);
    foo(aa);
```

```
    cout<<(*aa.ptr)<<endl;
    return 0;
}
```

Answer: The output of the code is an uncertain number, depending on the compiler used; for some compilers it could generate an error. The reason for this is that we do not define our own copy constructor.

When we call the foo function, the compiler will generate a default copy constructor which will shallow copy every data members defined in class A. This will lead to the result that two pointers, one in temporary object and the other in the object aa, will point to the same area in memory. When we get out the foo function, the compiler will automatically call the destructor function of the temporary object in which the pointer will be deleting and the area it points to will be free. In this situation, the pointer in aa will still point to the same area which has been free. When we try to visit the data through the pointer in aa, we will get garbage information.

Question 19. How do you overload an operator?

Answer:

```
type operator symbol (parameter_list);
```

If you define the operator outside of the class, then it will be a global operator function.

Example:

```
struct FooClass{int a;};
int operator + (FooClass lhs, FooClass rhs) {
    return lhs.a + rhs.a;
}
```

Question 20. What are smart pointers?

Answer: A smart pointer is a class built to mimic a pointer (offering dereferencing, indirection, arithmetic) that also offers extra features to simplify the usage, sharing and management of resources.

C++11 comes with three implementations of smart pointers: `shared_ptr`, `unique_ptr`, and `weak_ptr`.

Example:

```
//shared_pointer maintains
//a reference count
//when the count is zero the object
//pointed to is destroyed
std::shared_ptr<int> foo(new int(3));
std::shared_ptr<int> bar = foo;

//memory not released
//bar is still in scope
foo.reset();

//releases the memory,
//since no one is using it
bar.reset();
}
```

Question 21. What is encapsulation?

Answer: Encapsulation is the ability to expose an interface while hiding implementation. This is usually achieved through access modifiers (public, private, protected, etc.).

Question 22. What is a polymorphism?

Answer: Polymorphism is the ability for a set of classes to all be referenced through a common interface.

Question 23. What is inheritance?

Answer: Inheritance is the ability for one class to extend another through sub-classing. This is also referred to as "white-box" (the opposite of "black-box") re–use. A library can provide base classes that may be extended by the application developer.

Question 24. What is a virtual function? What is a pure virtual function and when do you use it?

Answer: Virtual functions are functions that are resolved by the compiler, at runtime, to the most derived version with the same signature. This means that if a function that was defined using a base class `Foo`, with a virtual member function `f`, is called using an instance of a sub class `FooChild`, that function is going to be dynamically binded to the implementation of the sub class (regardless that the actual code only refers to the base class).

A pure virtual function is a virtual function with no implementation in the base class, making the base class abstract (and thus can't be instantiated). Derived classes are forced to override the pure virtual function if they want to be instantiated. You use the same syntax as the virtual function but add `=0` to its declaration within the class.

Question 25. Why are virtual functions used for destructors? Can they be used for constructors?

Answer: Destructors are recommended to be defined as virtual, so the proper destructor (in the class hierarchy) is called at running time.

When calling a constructor, the caller needs to know the exact type of the object to be created, and thus they cannot be virtual.

Question 26. Write a function that computes the factorial of a positive integer.

Answer:

```
//for implementation
int factorial(int n){
    int output =1;
    for (int i =2 ; i <= n ; ++i)
        output *= i;
    return output;
}

//recursive implementation
int factorial(int n){
   if (n == 0) return 1;
   return n*factorial(n-1);
}

//tail recursive implementation
int factorial(int n, int last = 1){
    if (n == 0) return last;
    return factorial(n-1, last * n);
}
```

Question 27. Write a function that takes an array and returns the subarray with the largest sum.

Answer:

```
#include <vector>
#include <algorithm> // std::max

using namespace std;

template <typename T>
T max_sub_array(vector<T> const & numbers){
    T max_ending = 0, max_so_far = 0;
    for(auto & number: numbers){
        max_ending = max(0, max_ending + number);
```

```
        max_so_far = max(max_so_far, max_ending);
    }
    return max_so_far;
}
```

Question 28. Write a function that returns the prime factors of a positive integer.

Answer:

```
#include <vector>
using namepsace std;

vector<int> prime_factors(int n){
    vector<int> factors;
    for (int i = 2; i <= n/i ; ++i)
        while (n % i == 0) {
            factors.push_back(i);
            n /= i;
        }
    if (n > 1)
        factors.push_back(n);
    return factors;
}
```

Question 29. Write a function that takes a 64-bit integer and swaps the bits at indices `i` and `j`.

Answer:

```
long swap_bits (long x, const int &i, const int &j){
    if ( ((x >>i) & 1L) != ((x>>j) & 1L) )
        x ^= (1L <<i ) | (1L <<j);
    return x;
}
```

Question 30. Write a function that reverses a single linked list.

Answer:

```
#include <memory> // shared_ptr
using namespace std;

template<typename T>
struct node_t {
    T data;
    shared_ptr<node_T<T>> next;
};

//recursive implementation
template<typename T> shared_ptr<node_t<T>>
    reverse_linked_list(
        const shared_ptr<node_t <T>> &head){
    if (!head || !head->next) {
        return head;
    }

    shared_ptr<node_t<T>>
      new_head = reverse_linked_list(head->next);
    head->next->next = head;
    head->next = nullptr;
    return new_head;
}

//while implementation
template<typename T> shared_ptr<node_t<T>>
    reverse_linked_list(
        const shared_ptr<node_t <T>> &head){
    shared_ptr<node_t<T>>
        prev = nullptr, curr = head;
     while(curr) {
  shared_ptr<node_t<T>> temp = curr-> next;
  curr->next = prev;
```

```
  prev = curr;
  curr = temp;
     }
     return prev;
}
```

Question 31. Write a function that takes a string and returns true if its parenthesis are balanced.

Answer:

```
#include<string>
#include<stack>
using namespace std;

bool is_par_balanced(const string input)
{
    //"())())"=> false
    //"(a(dd)()(()))"=>true
    stack<char> par_stack;
    for(auto &c: input)
    {
        if(c==')')
        {
            if(par_stack.empty())
            return false;
            else if(par_stack.top()=='(')
            par_stack.pop();
        }
        else if(c=='(')
        par_stack.push(c);
    }
    return par_stack.empty();
}
```

Question 32. Write a function that returns the height of an arbitrary binary tree.

Answer:

```
#include<memory> //std::shared_ptr
#include<algorithm> //std::max
using namespace std;

template <typename T>
struct BinaryTree {
    T data;
    shared_ptr<BinaryTree<T>> left, right;
};


template <typename T>
int height(
    const shared_ptr<BinaryTree<T>> &tree,
    int count = -1){
        if (!tree) return count;
        return max(
           height(tree->left, count + 1),
           height(tree->right, count +1));
}
```

6.5 Monte Carlo simulations. Numerical methods.

Question 1. How would you compute π using Monte Carlo simulations? What is the standard deviation of this method?

Answer: An Acceptance–Rejection type method can be used to approximate π as follows: generate N points uniformly in the $[-1,1] \times [-1,1]$ square and accept a point (x,y) if the point is in the unit disk $D(0,1)$, i.e., if $x^2 + y^2 \leq 1$. If the number of accepted points is A, then the ratio $\frac{A}{N}$ converges in the limit as $N \to \infty$ to the ratio of the area of the unit disk to the area of the square, which is equal to $\frac{\pi}{4}$. Therefore,

$$\pi \approx \frac{4A}{N}.$$

The standard deviation of the method is $O\left(\frac{1}{\sqrt{N}}\right)$. We make this more precise by computing below the coefficient of $\frac{1}{\sqrt{N}}$ in $O\left(\frac{1}{\sqrt{N}}\right)$; see (6.79).

Let $U_1, U_2, \cdots$ be a sequence of independent identically distributed bivariate random variables uniformly distributed in $[-1,1] \times [-1,1]$. Denote by $\mathbf{1}_{D(0,1)}$ the indicator function of the unit disk $D(0,1)$, i.e.,

$$\mathbf{1}_{D(0,1)}(x,y) = \begin{cases} 1, & \text{if } (x,y) \in D(0,1); \\ 0, & \text{otherwise.} \end{cases}$$

Let

$$X_i = \mathbf{1}_{D(0,1)}(U_i), \ \forall\, i \geq 1.$$

The random variables X_i, $i \geq 1$, are independent and

identically distributed, with

$$\begin{aligned} E[X_i] &= E\left[\mathbf{1}_{D(0,1)}(U_i)\right] = \iint\limits_{D(0,1)} \frac{1}{4} dx dy \\ &= \frac{\pi}{4}. \end{aligned} \tag{6.77}$$

Since X_i, $i \geq 1$, are integrable, it follows from the strong law of large numbers that

$$\lim_{n\to\infty} \frac{X_1 + X_2 + \cdots + X_n}{n} = \frac{\pi}{4} \quad \text{almost surely.}$$

Note that $X_1 + X_2 + \cdots + X_n$ counts how many points out of the randomly selected n points reside in the disk $D(0,1)$. Thus, for N large enough,

$$\frac{X_1 + X_2 + \cdots + X_N}{N} \approx \frac{\pi}{4}. \tag{6.78}$$

To calculate the variance of the estimation in (6.78), note that, for $1 \leq i \leq N$,

$$\left(\mathbf{1}_{D(0,1)}(U_i)\right)^2 = \mathbf{1}_{D(0,1)}(U_i), \ \forall\, 1 \leq i \leq N,$$

and therefore, using (6.77), we find that

$$\begin{aligned} \text{var}(X_i) &= E[X_i^2] - (E[X_i])^2 \\ &= E\left[\left(\mathbf{1}_{D(0,1)}(U_i)\right)^2\right] - \left(\frac{\pi}{4}\right)^2 \\ &= E\left[\mathbf{1}_{D(0,1)}(U_i)\right] - \frac{\pi^2}{16} \\ &= \frac{\pi}{4} - \frac{\pi^2}{16} \\ &= \frac{\pi(4-\pi)}{16}. \end{aligned}$$

Then,

$$\begin{aligned}
& \operatorname{var}\left(\frac{X_1 + X_2 + \cdots + X_N}{N}\right) \\
= & \frac{1}{N^2}\left(\operatorname{var}(X_1) + \cdots + \operatorname{var}(X_N)\right) \\
= & \frac{1}{N^2} \cdot N \frac{\pi(4-\pi)}{16} \\
= & \frac{\pi(4-\pi)}{16N}.
\end{aligned}$$

By taking the square root of the variance, we conclude that the standard deviation of this Monte Carlo method for estimating π is

$$\frac{\sqrt{\pi(4-\pi)}}{4} \frac{1}{\sqrt{N}} \approx 0.82 \frac{1}{\sqrt{N}}. \tag{6.79}$$

Question 2. What methods do you know for generating independent samples of the standard normal distribution?

Answer: The three most commonly used methods to generate independent standard normal samples are:

- Box–Muller (using the Marsaglia–Bray algorithm in order to avoid estimating trigonometric functions);
- Acceptance–Rejection;
- Inverse Transform.

For details on these methods, see Glasserman [2]. □

Question 3. How do you generate a geometric Brownian motion stock path using random numbers from a normal distribution?

Answer: Consider a stock whose price follows the geometric Brownian motion

$$dS_t = \mu S_t dt + \sigma S_t dW_t, \tag{6.80}$$

where μ and σ are the drift and the volatility of the stock price, and W_t is a Weiner process. To generate a price path for the stock between time 0 and time T, discretize the time interval into m equal time steps of size $\delta t = \frac{T}{m}$, and let $t_j = j\delta t$, for $j = 0 : m$.

By integrating (6.80) between t_j and t_{j+1}, it follows that

$$S_{t_{j+1}} - S_{t_j} = \mu \int_{t_j}^{t_{j+1}} S_t dt + \sigma \int_{t_j}^{t_{j+1}} S_t dW_t. \quad (6.81)$$

We use the following approximations:

$$\begin{aligned} \int_{t_j}^{t_{j+1}} S_t dt &\approx S_{t_j}(t_{j+1} - t_j) \\ &= S_{t_j}\delta t; \quad (6.82) \\ \int_{t_j}^{t_{j+1}} S_t dW_t &\approx S_{t_j}(W_{t_{j+1}} - W_{t_j}) \\ &= S_{t_j}\sqrt{\delta t}\, Z_{j+1}, \quad (6.83) \end{aligned}$$

where Z_{j+1} is a standard normal variable, since W_t is a Wiener process and therefore $W_{t_{j+1}} - W_{t_j}$ is a normal variable with mean 0 and variance $t_{j+1} - t_j = \delta t$, i.e.,

$$W_{t_{j+1}} - W_{t_j} = \sqrt{\delta t} Z_{j+1}. \quad (6.84)$$

Note that Z_j, for $j = 1 : m$, are independent standard normals.

If $z_1, z_2, \ldots, z_m$ are independent samples of the standard normal distribution, we obtain from (6.81–6.83) that (6.80) can be discretized as follows:

$$S_{t_{j+1}} - S_{t_j} = \mu S_{t_j}\delta t + \sigma S_{t_j}\sqrt{\delta t} z_{j+1},$$

for $j = 0 : (m-1)$, which can be written as

$$S_{t_{j+1}} = S_{t_j}\left(1 + \mu\delta t + \sigma\sqrt{\delta t} z_{j+1}\right),$$

for $j = 0 : (m-1)$.

Note that there is a very small probability that the price path above becomes negative, which is a drawback of using this discretization.

A price path which is always positive can be generated using Itô's formula to express (6.80) as

$$d(\ln(S_t)) \;=\; \left(\mu - \frac{\sigma^2}{2}\right) dt \;+\; \sigma dW_t. \qquad (6.85)$$

By integrating (6.85) between t_j and t_{j+1}, it follows that

$$\ln(S_{t_{j+1}}) - \ln(S_{t_j}) \;=\; \ln\left(\frac{S_{t_{j+1}}}{S_{t_j}}\right)$$

$$= \;\left(\mu - \frac{\sigma^2}{2}\right)(t_{j+1} - t_j) \;+\; \sigma(W_{t_{j+1}} - W_{t_j})$$

$$= \;\left(\mu - \frac{\sigma^2}{2}\right)\delta t \;+\; \sigma\sqrt{\delta t}Z_{j+1}, \qquad (6.86)$$

for $j = 0 : (m-1)$, where (6.84) was used for (6.86).

Then, (6.86) can be discretized as follows:

$$\ln\left(\frac{S_{t_{j+1}}}{S_{t_j}}\right) \;=\; \left(\mu - \frac{\sigma^2}{2}\right)\delta t + \sigma\sqrt{\delta t}z_{j+1},$$

for $j = 0 : (m-1)$, and therefore

$$S_{t_{j+1}} \;=\; S_{t_j}\exp\left(\left(\mu - \frac{\sigma^2}{2}\right)\delta t + \sigma\sqrt{\delta t}z_{j+1}\right),$$

for all $j = 0 : (m-1)$. □

Question 4. How do you generate a sample of the standard normal distribution from 12 independent samples of the uniform distribution on $[0,1]$?

Answer: If $u_1, u_2, \ldots, u_{12}$ are 12 independent samples of the uniform distribution on $[0,1]$, then

$$\sum_{i=1}^{12} u_i \;-\; 6 \qquad (6.87)$$

can be used as a sample of the standard normal distribution.

To see this, recall from the Central Limit Theorem that, if X_i, $i \geq 1$, are independent identically distributed random variables with finite expected value $E[X]$ and standard deviation $\sigma(X)$, then

$$\lim_{n \to \infty} \frac{\frac{1}{n}\left(\sum_{i=1}^{n} X_i\right) - E[X]}{\frac{\sigma(X)}{\sqrt{n}}} = Z, \tag{6.88}$$

where Z is the standard normal distribution and the convergence in (6.88) is in distribution.

Let $U_1, U_2, \ldots, U_{12}$ be 12 independent uniform distributions on $[0, 1]$. Using (6.88) for $n = 12$ and $X_i = U_i$, $i = 1 : 12$, we infer that

$$Z \approx \frac{\frac{1}{12}\left(\sum_{i=1}^{12} U_i\right) - E[U]}{\frac{\sigma(U)}{\sqrt{12}}}, \tag{6.89}$$

where

$$\begin{aligned} E[U] &= \int_0^1 u\, du = \frac{1}{2}; \qquad (6.90)\\ \sigma^2(U) &= E[U^2] - (E[U])^2 = \int_0^1 u^2 du - \left(\frac{1}{2}\right)^2 \\ &= \frac{1}{3} - \frac{1}{4} = \frac{1}{12}, \end{aligned}$$

and thus

$$\sigma(U) = \frac{1}{\sqrt{12}}. \tag{6.91}$$

From (6.89–6.91), we obtain that

$$\begin{aligned} Z &\approx \frac{\frac{1}{12}\left(\sum_{i=1}^{12} U_i\right) - \frac{1}{2}}{\frac{1}{12}} \\ &= \sum_{i=1}^{12} U_i - 6, \end{aligned}$$

and therefore (6.87) can be used as an approximate sample of the standard normal distribution.

Note that this is an inefficient method, since it uses 12 samples from the uniform distribution to generate one approximate sample of the standard normal distribution, and all the samples that it generates are in the interval $[-6, 6]$. More efficient methods for generating independent standard normal samples are Box–Muller, which uses two uniform distribution samples to generate two samples of the standard normal distribution, and Acceptance–Rejection. □

Question 5. What is the rate of convergence for Monte Carlo methods?

Answer: If n is the number of paths in the Monte Carlo simulation and m is the number of time steps between 0 and T used in the discretization of each path, then the convergence rate of the Monte Carlo simulation is

$$O\left(\max\left(\frac{T}{m}, \frac{1}{\sqrt{n}}\right)\right).$$

The estimate above holds for Monte Carlo simulations on multi asset derivative securities, i.e., is independent of the number of underlying assets of the derivative security, unlike finite differences and numerical integration methods, where the convergence slows down as the number of underlying assets increases. □

Question 6. What variance reduction techniques do you know?

Answer: The variance reduction techniques are used to reduce the constant factor corresponding to the Monte Carlo approximation error $O\left(\frac{1}{\sqrt{n}}\right)$. Some of the most commonly used variance reduction techniques are:

- Control Variates;

- Antithetic Variables;
- Moment Matching.

For details on these methods and their implementation, see Glasserman [2]. □

Question 7. How do you generate samples of normal random variables with correlation ρ?

Answer: Assume that you can generate two samples z_1 and z_2 from two independent standard normal variables Z_1 and Z_2. Let $X_1 = Z_1$ and $X_2 = \rho Z_1 + \sqrt{1-\rho^2} Z_2$. Note that X_2 is a linear combination of independent normal variables, and therefore X_2 is a normal variable as well. Also,

$$\begin{aligned} \operatorname{corr}(X_1, X_2) &= \operatorname{corr}(Z_1, \rho Z_1 + \sqrt{1-\rho^2} Z_2) \\ &= \rho \operatorname{corr}(Z_1, Z_1) \\ &\quad + \sqrt{1-\rho^2} \operatorname{corr}(Z_1, Z_2) \\ &= \rho \operatorname{var}(Z_1) \\ &= \rho, \end{aligned}$$

since $\operatorname{var}(Z_1) = 1$ and $\operatorname{corr}(Z_1, Z_2) = 0$, since Z_1 and Z_2 are independent.

We conclude that X_1 and X_2 are normal random variables with correlation ρ. Thus, starting with two independent standard normal samples z_1 and z_2,

$$x_1 = z_1 \quad \text{and} \quad x_2 = \rho z_1 + \sqrt{1-\rho^2} z_2$$

are samples of normal random variables with correlation ρ. □

Question 8. What is the order of convergence of the Newton's method?

Answer: If it is convergent, Newton's method is quadratically (second order) convergent.

Recall that, given an initial guess x_0, the Newton's method recursion for solving $f(x) = 0$, where $f : \mathbb{R} \to \mathbb{R}$, is

$$x_{k+1} = x_k - \frac{f(x_k)}{f'(x_k)}, \quad \forall\, k \geq 0. \tag{6.92}$$

The quadratic convergence of Newton's method can be stated formally as follows: Let x^* be a solution of $f(x) = 0$. If $f(x)$ is a twice differentiable function with $f''(x)$ continuous, if $f'(x^*) \neq 0$, and if x_0 is close enough to x^*, then there exists $M > 0$ and n_M a positive integer such that

$$\frac{|x_{k+1} - x^*|}{|x_k - x^*|^2} < M, \quad \forall\, k \geq n_M. \tag{6.93}$$

To provide the intuition behind (6.93), note that, since $f(x^*) = 0$, the recursion (6.92) can be written as

$$\begin{aligned} & x_{k+1} - x^* \\ = \ & x_k - x^* - \frac{f(x_k) - f(x^*)}{f'(x_k)} \\ = \ & \frac{f(x^*) - f(x_k) + (x_k - x^*) f'(x_k)}{f'(x_k)}. \end{aligned} \tag{6.94}$$

Recall from the linear Taylor expansion of the function $f(x)$ around the point x_k that, if $f''(x)$ is continuous, then there exists a point c_k between x^* and x_k such that

$$\begin{aligned} f(x^*) &= f(x_k) + (x^* - x_k) f'(x_k) \\ &\quad + \frac{(x^* - x_k)^2}{2} f''(c_k). \end{aligned} \tag{6.95}$$

From (6.94) and (6.95), we find that

$$x_{k+1} - x^* = (x^* - x_k)^2 \frac{f''(c_k)}{2f'(x_k)},$$

and conclude that

$$\frac{|x_{k+1} - x^*|}{|x_k - x^*|^2} = \left| \frac{f''(c_k)}{2f'(x_k)} \right|. \tag{6.96}$$

If $f''(x)$ and $f'(x)$ are continuous functions such that $f'(x^*) \neq 0$, it follows that, if x_k is close to x^*, then $\left|\frac{f''(c_k)}{2f'(x_k)}\right|$ is close to $\left|\frac{f''(x^*)}{2f'(x^*)}\right| < \infty$. Therefore, the term $\left|\frac{f''(c_k)}{2f'(x_k)}\right|$ is uniformly bounded if x_k is close enough to x^*, and (6.96) can be written formally as (6.93). □

Question 9. Which finite difference method corresponds to trinomial trees?

Answer: Forward Euler. As an explicit finite difference method, the Forward Euler discretization of the Black–Scholes PDE gives the finite difference value of the option at a node as a linear combination of the option values at three nodes on the prior time step, which is similar to the risk neutral formula for trinomial trees.

If the calibration of the trinomial trees is done in the log space, i.e., if the up and down factors are calibrated to the normal distribution of $\ln(S)$, and if the Forward Euler discretization is done for the constant coefficients PDE obtained by the change of variables $x = \ln(S)$, the classical trinomial tree recursion and the Forward Euler recursion are almost identical. □

6.6 Probability. Stochastic calculus.

Question 1. What is the exponential distribution? What are the mean and the variance of the exponential distribution?

Answer: The density function of the exponential random variable X with parameter $\alpha > 0$ is

$$f(x) = \begin{cases} \alpha\, e^{-\alpha x}, & \text{if } x \geq 0; \\ 0, & \text{if } x < 0. \end{cases}$$

The expected value and the variance of the exponential random variable X are

$$E[X] = \frac{1}{\alpha} \quad \text{and} \quad \mathrm{var}(X) = \frac{1}{\alpha^2}.$$

To see this, use integration by parts to find that

$$\begin{aligned} \int_0^\infty x e^{-\alpha x}\, dx &= -\frac{x e^{-\alpha x}}{\alpha}\Big|_0^\infty + \frac{1}{\alpha}\int_0^\infty e^{-\alpha x}\, dx \\ &= 0 - \frac{e^{-\alpha x}}{\alpha^2}\Big|_0^\infty = \frac{1}{\alpha^2}; \\ \int_0^\infty x^2 e^{-\alpha x}\, dx &= -\frac{x^2 e^{-\alpha x}}{\alpha}\Big|_0^\infty + \frac{2}{\alpha}\int_0^\infty x e^{-\alpha x}\, dx \\ &= 0 + \frac{2}{\alpha} \times \frac{1}{\alpha^2} = \frac{2}{\alpha^3}. \end{aligned}$$

Then,

$$E[X] = \int_{-\infty}^\infty x f(x)\, dx = \alpha \int_0^\infty x e^{-\alpha x}\, dx = \frac{1}{\alpha};$$

$$E[X^2] = \int_{-\infty}^\infty x^2 f(x)\, dx = \alpha \int_0^\infty x^2 e^{-\alpha x}\, dx = \frac{2}{\alpha^2}.$$

Therefore,

$$\mathrm{var}(X) = E[X^2] - (E[X])^2 = \frac{2}{\alpha^2} - \left(\frac{1}{\alpha}\right)^2 = \frac{1}{\alpha^2}. \quad \square$$

Question 2. If X and Y are independent exponential random variables with mean 6 and 8, respectively, what is the probability that Y is greater than X?

Answer: The probability density functions of X and Y are, respectively:

$$f_X(x) = \begin{cases} \frac{1}{6}e^{-\frac{x}{6}}, & \text{if } x \geq 0; \\ \\ 0, & \text{if } x < 0; \end{cases}$$

$$f_Y(y) = \begin{cases} \frac{1}{8}e^{-\frac{y}{8}}, & \text{if } y \geq 0; \\ \\ 0, & \text{if } y < 0. \end{cases}$$

Since X and Y are independent, the joint probability density function $f_{XY}(x, y)$ of (X, Y) is the product of the marginal probability density functions, i.e.,

$$\begin{aligned} f_{XY}(x, y) &= f_X(x)\, f_Y(y) \\ &= \begin{cases} \frac{1}{48}e^{-\frac{x}{6}-\frac{y}{8}}, & \text{if } x \geq 0,\ y \geq 0; \\ \\ 0, & \text{otherwise.} \end{cases} \end{aligned}$$

Let

$$A = \{(x, y) \in \mathbb{R}^2 : y \geq x\}.$$

The probability that Y is greater than X can be found by evaluating the double integral

$$\begin{aligned} P(Y \geq X) &= \iint_A f_{XY}(x, y)dxdy \\ &= \frac{1}{48}\int_0^\infty \int_0^y e^{-\frac{x}{6}-\frac{y}{8}}\, dxdy \end{aligned}$$

as follows:

$$\begin{aligned} P(Y \geq X) &= \frac{1}{48} \int_0^\infty e^{-\frac{y}{8}} \left(\int_0^y e^{-\frac{x}{6}} \, dx \right) dy \\ &= \frac{1}{48} \int_0^\infty e^{-\frac{y}{8}} \left(-6e^{-\frac{x}{6}} \Big|_0^y \right) dy \\ &= \frac{1}{8} \int_0^\infty e^{-\frac{y}{8}} \left(1 - e^{-\frac{y}{6}} \right) dy \\ &= \frac{1}{8} \int_0^\infty \left(e^{-\frac{y}{8}} - e^{-\frac{7}{24}y} \right) dy \\ &= \frac{1}{8} \left(8 - \frac{24}{7} \right) \\ &= \frac{4}{7}. \quad \square \end{aligned}$$

Question 3. What are the expected value and the variance of the Poisson distribution?

Answer: A Poisson distribution is a random variable X taking nonnegative integer values with probabilities

$$P(X = k) = \frac{e^{-\lambda} \lambda^k}{k!}, \quad \forall \, k \geq 0,$$

where $\lambda > 0$ is a fixed positive number.

We show that the expected value and the variance of the Poisson distribution X are

$$E[X] = \lambda \quad \text{and} \quad \text{var}(X) = \lambda.$$

By definition,

$$\begin{aligned} E[X] &= \sum_{k=0}^{\infty} P(X = k) \cdot k = \sum_{k=1}^{\infty} \frac{e^{-\lambda} \lambda^k}{k!} k \\ &= e^{-\lambda} \lambda \sum_{k=1}^{\infty} \frac{\lambda^{k-1}}{(k-1)!}. \end{aligned} \tag{6.97}$$

Since the Taylor series expansion for e^t is

$$e^t = \sum_{k=0}^{\infty} \frac{t^k}{k!},$$

it follows that

$$\sum_{k=1}^{\infty} \frac{\lambda^{k-1}}{(k-1)!} = e^{\lambda}; \tag{6.98}$$

$$\sum_{k=2}^{\infty} \frac{\lambda^{k-2}}{(k-2)!} = e^{\lambda}. \tag{6.99}$$

From (6.97) and (6.98), we find that $E[X] = \lambda$.

To calculate $\mathrm{var}(X)$, note that

$$\begin{aligned} E[X^2] &= \sum_{k=0}^{\infty} P(X=k) \cdot k^2 \\ &= \sum_{k=1}^{\infty} \frac{e^{-\lambda}\lambda^k}{k!} k^2 \\ &= e^{-\lambda} \sum_{k=1}^{\infty} \frac{k\lambda^k}{(k-1)!}, \end{aligned}$$

which can be written as

$$\begin{aligned} E[X^2] &= e^{-\lambda} \sum_{k=1}^{\infty} \frac{(k-1)\lambda^k}{(k-1)!} + e^{-\lambda} \sum_{k=1}^{\infty} \frac{\lambda^k}{(k-1)!} \\ &= e^{-\lambda}\lambda^2 \sum_{k=2}^{\infty} \frac{\lambda^{k-2}}{(k-2)!} + e^{-\lambda}\lambda \sum_{k=1}^{\infty} \frac{\lambda^{k-1}}{(k-1)!} \\ &= \lambda^2 + \lambda, \end{aligned}$$

where (6.98) and (6.99) were used for the last equality.

We conclude that

$$\mathrm{var}(X) = E[X^2] - (E[X])^2 = \lambda. \quad \square$$

Question 4. A point is chosen uniformly from the unit disk. What is the expected value of the distance between the point and the center of the disk?

Answer: The expected value of the distance between a uniformly chosen point in the unit disk D and the center of the disk can be computed as $E\left[\sqrt{X^2+Y^2}\right]$, where (X,Y) is uniformly distributed in the unit disk D. The probability density function of (X,Y) is

$$f(x,y) = \begin{cases} \frac{1}{\pi}, & \text{if } x \in D; \\ 0, & \text{otherwise.} \end{cases}$$

Then,

$$E\left[\sqrt{X^2+Y^2}\right] = \frac{1}{\pi}\iint_D \sqrt{x^2+y^2}\,dxdy. \qquad (6.100)$$

Using the polar coordinates substitution $x = r\cos\theta$ and $y = r\sin\theta$, with $0 \le r \le 1$ and $0 \le \theta < 2\pi$, and recalling that $dxdy = rdrd\theta$, we obtain from (6.100) that

$$\begin{aligned} & E\left[\sqrt{X^2+Y^2}\right] \\ = \ & \frac{1}{\pi}\int_0^{2\pi}\int_0^1 \sqrt{r^2\left(\cos^2\theta+\sin^2\theta\right)}\,rdrd\theta \\ = \ & \frac{1}{\pi}\int_0^1\int_0^{2\pi} r^2\,d\theta\,dr \qquad (6.101) \\ = \ & \frac{1}{\pi}\int_0^1 2\pi r^2\,dr = 2\int_0^1 r^2 dr \\ = \ & \frac{2}{3}, \end{aligned}$$

where for (6.101) we used the fact that $\cos^2\theta + \sin^2\theta = 1$ for all θ. $\square$

Question 5. Consider two random variables X and Y with mean 0 and variance 1, and with joint normal distribution. If $\operatorname{cov}(X,Y) = \frac{1}{\sqrt{2}}$, what is the conditional probability $P(X > 0|Y < 0)$?

Answer: From the definition of conditional probability, it follows that

$$P(X > 0|Y < 0) = \frac{P(X > 0, Y < 0)}{P(Y < 0)}. \tag{6.102}$$

Note that

$$P(Y < 0) = \frac{1}{2}, \tag{6.103}$$

since Y is a standard normal random variable.

In order to compute $P(X > 0, Y < 0)$, let

$$W = \sqrt{2}\,X - Y. \tag{6.104}$$

Since $E[X] = E[Y] = 0$, it follows that $E[W] = 0$. Moreover, since

$$\operatorname{var}(X) = \operatorname{var}(Y) = 1 \text{ and } \operatorname{cov}(X,Y) = \frac{1}{\sqrt{2}},$$

we obtain that

$$\begin{aligned} \operatorname{var}(W) &= \operatorname{var}\left(\sqrt{2}\,X - Y\right) \\ &= \operatorname{var}\left(\sqrt{2}\,X\right) - 2\operatorname{cov}\left(\sqrt{2}X, Y\right) + \operatorname{var}(Y) \\ &= 2\operatorname{var}(X) - 2\sqrt{2}\operatorname{cov}(X,Y) + \operatorname{var}(Y) \\ &= 1, \end{aligned}$$

and

$$\begin{aligned} \operatorname{cov}(W,Y) &= \operatorname{cov}\left(\sqrt{2}\,X - Y, Y\right) \\ &= \sqrt{2}\operatorname{cov}(X,Y) - \operatorname{var}(Y) \\ &= 0. \end{aligned}$$

Note that $W = \sqrt{2}\,X - Y$ is a normal random variable since X and Y have joint normal distribution. Moreover, since $E[W] = 0$, $\text{var}(W) = 1$, and $\text{cov}(W, Y) = 0$, it follows that W and Y are independent standard normal variables.

From (6.104), we find that

$$X = \frac{1}{\sqrt{2}}(W + Y).$$

Then, the probability of the event $\{X > 0, Y < 0\}$ can be written as

$$\begin{aligned} & P(X > 0, Y < 0) \\ = \; & P\left(\frac{1}{\sqrt{2}}(W + Y) > 0, Y < 0\right) \\ = \; & P(W + Y > 0, Y < 0). \end{aligned} \quad (6.105)$$

The two straight lines $w + y = 0$ and $y = 0$ cut the (w, y) plane into four wedges:

$$\begin{aligned} R_1 &= \{w + y > 0, y < 0\}; \\ R_2 &= \{w + y > 0, y > 0\}; \\ R_3 &= \{w + y < 0, y < 0\}; \\ R_4 &= \{w + y < 0, y > 0\}. \end{aligned}$$

Note that

$$P(W + Y > 0, Y < 0) = P((W, Y) \in R_1). \quad (6.106)$$

Since W and Y are independent normal random variables, their joint probability density function is rotationally symmetric, and therefore

$$\begin{aligned} P((W, Y) \in R_1) &= P((W, Y) \in R_4); \quad (6.107) \\ P((W, Y) \in R_2) &= P((W, Y) \in R_3); \quad (6.108) \\ P((W, Y) \in R_2) &= 3P((W, Y) \in R_1); \end{aligned}$$

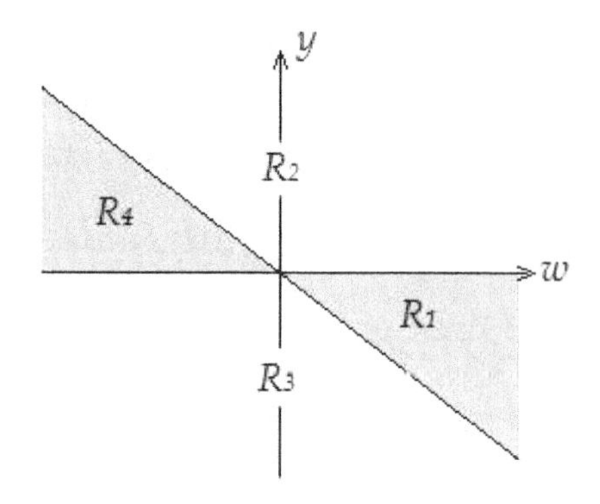

图 6.1: The regions R_1 to R_4 in the (w, y) plane.

see Figure 6.1.

Also, note that

$$\sum_{i=1}^{4} P((W, Y) \in R_i) = 1, \tag{6.109}$$

since $P(W + Y = 0 \text{ or } Y = 0) = 0$.

From (6.107–6.109), we find that

$$\begin{aligned} 1 &= \sum_{i=1}^{4} P((W, Y) \in R_i) \\ &= 2P((W, Y) \in R_1) + 2P((W, Y) \in R_2) \\ &= 8P((W, Y) \in R_1), \end{aligned}$$

and therefore

$$P((W, Y) \in R_1) = \frac{1}{8}.$$

Thus,

$$P(W + Y > 0, Y < 0) = \frac{1}{8}; \tag{6.110}$$

see (6.106).

Then, from (6.105) and (6.110), it follows that

$$P(X > 0, Y < 0) = \frac{1}{8}. \qquad (6.111)$$

From (6.102), (6.103), and (6.111), we conclude that

$$\begin{aligned} P(X > 0 | Y < 0) &= \frac{P(X > 0, Y < 0)}{P(Y < 0)} \\ &= \frac{1}{4}. \quad \square \end{aligned}$$

Question 6. If X and Y are lognormal random variables, is their product XY lognormally distributed?

Answer: First, note that, if X and Y are independent lognormal random variables, then XY is lognormally distributed, since $\ln(XY) = \ln(X) + \ln(Y)$ is the sum of two independent normal random variables, and therefore it is normally distributed.

In a more general setting, if $\ln(X)$ and $\ln(Y)$ have joint normal distribution, then $\ln(X) + \ln(Y)$ is normally distributed and therefore XY is a lognormal random variable.

Otherwise, $\ln(X) + \ln(Y)$ may not be normally distributed even if $\ln(X)$ and $\ln(Y)$ are normally distributed, in which case XY is not lognormally distributed. $\square$

Question 7. Let X be a normal random variable with mean μ and variance σ^2, and let Φ be the cumulative distribution function of the standard normal distribution. Find the expected value of $Y = \Phi(X)$.

Answer: Let Z be a standard normal random variable independent of X. Then,

$$Y = \Phi(X) = P(Z \leq X | X) = E[\mathbf{1}_{Z \leq X} | X],$$

and therefore

$$E[Y] = E\left[E\left[\mathbf{1}_{Z\leq X} \mid X\right]\right]. \qquad (6.112)$$

Recall from the Tower Property for conditional expectation[10] that, for any two random variables T and W,

$$E[T] = E[E[T|W]]. \qquad (6.113)$$

Using (6.113) for $T = \mathbf{1}_{Z\leq X}$ and $W = X$, we obtain that

$$\begin{aligned} E\left[E\left[\mathbf{1}_{Z\leq X} \mid X\right]\right] &= E\left[\mathbf{1}_{Z\leq X}\right] \\ &= P(Z \leq X). \end{aligned} \qquad (6.114)$$

From (6.112) and (6.114), we obtain that

$$E[Y] = P(Z \leq X). \qquad (6.115)$$

Recall that X is a normal random variable with mean μ and variance σ^2, and that X and Z are independent. Then, $Z - X$ is a normal random variable with the following mean and variance:

$$\begin{aligned} E[Z - X] &= -\mu; \\ \mathrm{var}(Z - X) &= \mathrm{var}(Z) + \mathrm{var}(X) = 1 + \sigma^2. \end{aligned}$$

and $Z - X$ is a normal random variable with mean $-\mu$ and variance $1 + \sigma^2$. Thus, $\frac{Z-X+\mu}{\sqrt{1+\sigma^2}}$ is a standard normal random variable and therefore

$$\begin{aligned} P(Z \leq X) &= P(Z - X \leq 0) \\ &= P\left(\frac{Z - X + \mu}{\sqrt{1+\sigma^2}} \leq \frac{\mu}{\sqrt{1+\sigma^2}}\right) \\ &= \Phi\left(\frac{\mu}{\sqrt{1+\sigma^2}}\right). \end{aligned} \qquad (6.116)$$

[10]In other words, to calculate the expected value of T, one can first calculate the conditional expected value of T knowing the extra information from W, then average out the resulting conditional expected value over W.

From (6.115) and (6.116) we conclude that

$$E[Y] = \Phi\left(\frac{\mu}{\sqrt{1+\sigma^2}}\right). \qquad (6.117)$$

We note that, if X is the standard normal variable, then $Y = \Phi(X)$ is uniformly distributed in the interval $[0, 1]$, and therefore $E[Y] = \frac{1}{2}$. Note that, for $\mu = 0$ and $\sigma = 1$ in (6.117), i.e., if X is a standard normal variable, then $E[Y] = \Phi(0) = \frac{1}{2}$, which is consistent to the comment above. □

Question 8. What is the law of large numbers?

Answer: There is a strong law of large numbers and there is a weak law of large numbers. The strong law of large numbers states that the average of a large number of independent identically distributed integrable random variables converges almost surely to their common mean; in the case of the weak law of large numbers, the convergence is only in probability.

More precisely, let $X_1, X_2, \ldots$ be a sequence of independent identically distributed random variables with finite expected value $\mu = E[X_i]$, and let $S_n = X_1 + \cdots + X_n$.

The strong law of large numbers states that $\frac{S_n}{n} \to \mu$ almost surely, i.e.,

$$P\left(\lim_{n\to\infty} \frac{S_n}{n} = \mu\right) = 1. \qquad (6.118)$$

The weak law of large numbers states that $\frac{S_n}{n} \to \mu$ in probability, i.e.,

$$\lim_{n\to\infty} P\left(\left|\frac{S_n}{n} - \mu\right| > \epsilon\right) = 0, \quad \forall\, \epsilon > 0. \qquad (6.119)$$

Note that, if a sequence of random variables convergences almost surely, than it also converges in probability, and therefore, if (6.118) holds true, then (6.119) also holds true. This is the sense in which the strong law of large numbers is "stronger" than the weak law of large numbers. □

Question 9. What is the central limit theorem?

Answer: The central limit theorem states that the limiting distribution of the centered and scaled sum of an independent identically distributed sequence of random variables is a normal distribution if the common distribution of the random variables has finite variance.

More precisely, let X_1, X_2, ... be a sequence of independent identically distributed random variables with finite expected value $\mu = E[X_i]$ and finite variance $\sigma^2 = \mathrm{var}[X_i]$. Let $S_n = X_1 + \cdots + X_n$. Then,

$$\lim_{n\to\infty} \frac{S_n - n\mu}{\sigma\sqrt{n}} = Z,$$

where Z is the standard normal distribution, and the convergence is in distribution, i.e.,

$$\lim_{n\to\infty} P\left(\frac{S_n - n\mu}{\sigma\sqrt{n}} \le t\right) = P(Z \le t).$$

Putting together the Law of Large Numbers and the Central Limit Theorem, the following approximation as n goes to infinity holds:

$$\frac{S_n}{n} \approx \mu + \frac{\sigma}{\sqrt{n}} Z. \quad \square$$

Question 10. What is a martingale? How is it related to option pricing?

Answer: Let $(\Omega, \mathcal{F}_t, P)$ be a filtered probability space, where Ω is the sample space, $\mathcal{F}_t$ is a filtration, and P is a probability measure on Ω. A stochastic process X_t

is called a *martingale* with respect to the filtration $\{\mathcal{F}_t\}$ if and only if
(i) X_t is adapted, i.e., X_t is $\mathcal{F}_t$-measurable for all t;
(ii) X_t is integrable for all t, i.e., $E[|X_t|] < \infty$ for all t;
(iii) $E[X_t|\mathcal{F}_s] = X_s$ almost surely for all $s < t$.

In other words, a martingale is a stochastic process in which, given the available information $\mathcal{F}_s$ up to the current time s, the optimal estimation (in the least square sense) of the process in the future time t, i.e., $E[X_t|\mathcal{F}_s]$, is the current value X_s almost surely.

The martingale concept is one of the cornerstones of option pricing theory. The fundamental theorem of asset pricing states that, if a market model is arbitrage–free, then there exists a risk neutral probability so that each discounted asset price process under the risk neutral probability is a martingale. Thus, a way to price a derivative security is to figure out a partial differential equation, called the pricing equation, usually deduced by applying Itô's formula, so that the discounted price process of the derivative is a martingale under risk neutral probability.

Question 11. Explain the assumption $(dW_t)^2 = dt$ used in the informal derivation of Itô's Lemma.

Answer: The notation in differential form $(dW_t)^2 = dt$ is a shorthand for the conventional integral notation used in Riemann integral

$$\int_0^T (dW_t)^2 = \int_0^T dt. \tag{6.120}$$

The intuition behind (6.120) is related to the *quadratic variation* of a Brownian path. By definition, the quadratic variation $QV_W[0,T]$ of the Brownian path W in the interval $[0,T]$ is

$$QV_W[0,T] = \lim_{n\to\infty} \sum_{i=1}^{n} \left|W_{t_i} - W_{t_{i-1}}\right|^2,$$

where $t_i = \frac{i}{n}T$, for $i = 0 : n$. Therefore, in expectation, we have that

$$\begin{aligned} & E\left[\lim_{n\to\infty}\sum_{i=1}^{n}\left|W_{t_i} - W_{t_{i-1}}\right|^2\right] \\ = & \lim_{n\to\infty}\sum_{i=1}^{n} E\left[\left|W_{t_i} - W_{t_{i-1}}\right|^2\right] \\ = & \lim_{n\to\infty}\sum_{i=1}^{n}(t_i - t_{i-1}) \\ = & T, \end{aligned}$$

since $W_{t_i} - W_{t_{i-1}} \sim N(0, t_i - t_{i-1})$ for $i = 1 : n$. In fact, it can be shown that the convergence

$$\sum_{i=1}^{n}\left|W_{t_i} - W_{t_{i-1}}\right|^2 \longrightarrow T \text{ as } n \to \infty$$

is in L^2 sense, i.e.,

$$\lim_{n\to\infty} E\left[\left|\sum_{i=1}^{n}\left|W_{t_i} - W_{t_{i-1}}\right|^2 - T\right|^2\right] = 0.$$

By mimicking the notations used in the conventional Riemann integral, we write that

$$\begin{aligned} \int_0^T (dW_t)^2 &= \lim_{n\to\infty}\sum_{i=1}^{n}\left|W_{t_i} - W_{t_{i-1}}\right|^2 = T \\ &= \int_0^T dt, \end{aligned}$$

whose shorthand notation in differential form is

$$(dW_t)^2 = dt. \quad \square$$

Question 12. If W_t is a Wiener process, find $E[W_t W_s]$.

Answer: Assume that $s \leq t$. Write W_t as $W_t = (W_t - W_s) + W_s$, and note that

$$W_t W_s = (W_t - W_s)W_s + W_s^2. \tag{6.121}$$

Since W_τ, $\tau \geq 0$, is a Wiener process, it follows that $W_t - W_s$ and W_s are independent normal random variables of mean 0, and therefore

$$\begin{aligned} E[(W_t - W_s)W_s] &= E[W_t - W_s]E[W_s] \\ &= 0. \end{aligned} \tag{6.122}$$

Also, W_s is normal of mean $E[W_s] = 0$ and variance $\mathrm{var}(W_s) = s$. Thus,

$$\begin{aligned} \mathrm{var}(W_s) &= E[W_s^2] - (E[W_s])^2 \\ &= E[W_s^2], \end{aligned}$$

and therefore

$$E[W_s^2] = \mathrm{var}(W_s) = s. \tag{6.123}$$

From (6.121–6.123), we obtain that

$$\begin{aligned} E[W_t W_s] &= E[(W_t - W_s)W_s] + E[W_s^2] \\ &= s. \end{aligned} \tag{6.124}$$

Since (6.124) was derived under the assumption $s \leq t$, we conclude that

$$E[W_t W_s] = \min(s, t). \quad \square \tag{6.125}$$

Question 13. If W_t is a Wiener process, what is $\mathrm{var}(W_t + W_s)$?

Answer: Assume that $s \leq t$. Write W_t as $W_t = (W_t - W_s) + W_s$, and note that $W_t + W_s = (W_t - W_s) + 2W_s$.

Then,

$$\begin{aligned} \mathrm{var}(W_t + W_s) &= \mathrm{var}(W_t - W_s) \\ &\quad + 4\mathrm{cov}(W_t - W_s, W_s) \\ &\quad + 4\mathrm{var}(W_s). \end{aligned} \tag{6.126}$$

Since W_τ, $\tau \geq 0$, is a Wiener process, it follows that

$$\mathrm{cov}(W_t - W_s, W_s) = 0; \tag{6.127}$$

$$\mathrm{var}(W_t - W_s) = t - s; \tag{6.128}$$

$$\mathrm{var}(W_s) = s, \tag{6.129}$$

since $W_t - W_s$ and W_s are independent normal variables of variance $t - s$ and s, respectively.

From (6.126–6.129), we obtain that

$$\mathrm{var}(W_t + W_s) = (t - s) + 4s = t + 3s. \tag{6.130}$$

Since (6.130) was derived under the assumption $s \leq t$, we conclude that

$$\mathrm{var}(W_t + W_s) = \max(s, t) + 3\min(s, t). \quad \square$$

Question 14. Let W_t be a Wiener process. Find Find

$$\int_0^t W_s \, dW_s \quad \text{and} \quad E\left[\int_0^t W_s \, dW_s \right].$$

Answer: Since $\int x \, dx = \frac{x^2}{2} + C$, we begin by computing $d\left(\frac{W_t^2}{2}\right)$. Recall from Itô's lemma that, if $f(x,t)$ is a continuously differentiable function, then

$$df = \frac{\partial f}{\partial t} \, dt + \frac{\partial f}{\partial x} \, dW_t + \frac{1}{2} \frac{\partial^2 f}{\partial x^2} \, dt.$$

For $f(x,t) = \frac{x^2}{2}$, we find that

$$d\left(\frac{W_t^2}{2}\right) = W_t dW_t + \frac{1}{2} dt,$$

and therefore

$$d\left(\frac{W_t^2 - t}{2}\right) = W_t dW_t. \qquad (6.131)$$

By integrating (6.131) between 0 and t, and since $W_0 = 0$, we obtain that

$$\int_0^t W_s \, dW_s = \frac{W_t^2 - t}{2}, \quad \forall\, t \geq 0. \qquad (6.132)$$

From (6.132), it follows that

$$E\left[\int_0^t W_s \, dW_s\right] = \frac{E[W_t^2] - t}{2} = 0,$$

since W_t is normally distributed with mean 0 and variance t and therefore

$$\begin{aligned} E[W_t^2] &= \mathrm{var}(W_t) + (E[W_t])^2 \\ &= t. \quad \square \end{aligned}$$

Question 15. Find the distribution of the random variable

$$X = \int_0^1 W_t dW_t.$$

Answer: Recall from Itô's formula that, if W_t is a Wiener process and $f(x)$ is a function with continuous second order derivative, then

$$df(W_t) = f'(W_t)dW_t + \frac{1}{2}f''(W_t)dt. \qquad (6.133)$$

For $f(x) = x^2$, we obtain from (6.133) that

$$dW_t^2 = 2W_t dW_t + dt. \qquad (6.134)$$

By integrating (6.134) from 0 to 1, we obtain that

$$W_1^2 = 2\int_0^1 W_t dW_t + 1 = 2X + 1. \qquad (6.135)$$

Note that W_1 is a standard normal random variable, and let $W_1 = Z$. By solving (6.135) for X, we find that

$$X = \frac{W_1^2 - 1}{2} = \frac{Z^2 - 1}{2}.$$

Let $f_X(x)$ and $F_X(x)$ be the probability density function of X and the cumulative distribution function of X, respectively.

Note that

$$P\left(X \leq -\frac{1}{2}\right) = P(Z^2 \leq 0) = 0.$$

Thus, if $x < -\frac{1}{2}$, then $F_X(x) = P(X \leq x) = 0$ and therefore

$$f_X(x) = 0,$$

since $f_X(x) = F_X'(x)$.

If $x \geq -\frac{1}{2}$, then

$$\begin{aligned} F_X(x) &= P(X \leq x) = P\left(\frac{Z^2 - 1}{2} \leq x\right) \\ &= P\left(Z^2 \leq 2x + 1\right) \\ &= P\left(-\sqrt{2x+1} \leq Z \leq \sqrt{2x+1}\right) \\ &= 2\,P\left(0 \leq Z \leq \sqrt{2x+1}\right) \\ &= \frac{2}{\sqrt{2\pi}} \int_0^{\sqrt{2x+1}} e^{-\frac{y^2}{2}}\, dy. \end{aligned}$$

By differentiating $F_X(x)$, it follows that, for $x < -\frac{1}{2}$,

$$\begin{aligned} f_X(x) &= F_X'(x) \\ &= \sqrt{\frac{2}{\pi}} \frac{d}{dx} \left(\int_0^{\sqrt{2x+1}} e^{-\frac{y^2}{2}} dy \right) \\ &= \sqrt{\frac{2}{\pi}} e^{-\frac{(\sqrt{2x+1})^2}{2}} \left(\sqrt{2x+1}\right)' \\ &= \sqrt{\frac{2}{\pi}} \frac{e^{-\frac{2x+1}{2}}}{\sqrt{2x+1}}. \end{aligned}$$

We conclude that

$$f_X(x) = \begin{cases} \sqrt{\frac{2}{\pi}} \frac{e^{-\frac{2x+1}{2}}}{\sqrt{2x+1}}, & \text{if } x > -\frac{1}{2}; \\ 0 & \text{if } x \leq -\frac{1}{2}. \end{cases} \qquad \square$$

Question 16. Let W_t be a Wiener process. Find the mean and the variance of

$$\int_0^t W_s^2 dW_s.$$

Answer: Let $\mathcal{B}_t$ be the Borel σ-algebra over the time interval $[0, t]$ and let $\mathcal{F}_t$ be the filtration for the probability space in which the Wiener process W_t resides. We will use the following results:[11]

Theorem 6.1. *(Martingality)*
Let f_s be progressively measurable and square integrable in $[0, T]$, i.e., f_t is $\mathcal{B}_t \otimes \mathcal{F}_t$-measurable for every $t \in [0, T]$ and $E\left[\int_0^T |f_t|^2 dt\right] < \infty$. Then, the stochastic integral

[11] For the proofs of Theorem 6.1 and Theorem 6.2, see Theorem 2.8 on p.65 and Theorem 3.1 on p.67, respectively, from Friedman [1].

$\int_0^t f_s dW_s$ defines a zero mean, square integrable martingale for $t \in [0,T]$. In particular,

$$E\left[\int_0^t f_s dW_s\right] = 0, \quad \forall\, t \in [0,T]. \tag{6.136}$$

Theorem 6.2. *(Itô's isometry)*
Let f_t and g_t be progressively measurable and square integrable processes. Then,

$$E\left[\int_0^t f_s dW_s \int_0^t g_s dW_s\right] = \int_0^t E[f_s\, g_s] ds.$$

In particular, if $f_s = g_s$, it follows that

$$E\left[\left(\int_0^t f_s dW_s\right)^2\right] = \int_0^t E[f_s^2] ds. \tag{6.137}$$

For our problem, we need to compute $E[X]$ and $\mathrm{var}(X)$, where

$$X = \int_0^t W_s^2 dW_s.$$

We first check that the integrand W_s^2 is progressively measurable and square integrable in $[0,t]$.

Note that W_s^2 is progressively measurable because it is adapted and continuous.

Furthermore, since $W_s \sim N(0,s)$, it follows that $W_s = \sqrt{s}Z$, where Z is a standard normal random variable, and therefore

$$E[W_s^4] = E\left[\left(\sqrt{s}Z\right)^4\right] = s^2\, E[Z^4] = 3s^2, \tag{6.138}$$

where we used the fact that the fourth moment of the standard normal distribution is 3, i.e., $E[Z^4] = 3$. From (6.138), it follows that

$$\int_0^t E[W_s^4] ds = \int_0^t 3s^2 ds = t^3 < \infty. \tag{6.139}$$

In other words, W_s^2 is square integrable in $[0, t]$.

We can therefore apply both Theorems 6.1 and 6.2 with $f_s = W_s^2$ and $g_s = W_s^2$.

From (6.136) for $f_s = W_s^2$, we find that

$$E[X] = E\left[\int_0^t W_s^2 dW_s\right] = 0. \tag{6.140}$$

From (6.137) for $f_s = g_s = W_s^2$, and using (6.139), it follows that

$$\begin{aligned} E[X^2] &= E\left[\left(\int_0^t W_s^2 dW_s\right)^2\right] = \int_0^t E[W_s^4] ds \\ &= t^3. \end{aligned}$$

Since $E[X] = 0$, see (6.140), we find that

$$\mathrm{var}(X) = E[X^2] - (E[X])^2 = t^3. \quad \square$$

Question 17. If W_t is a Wiener process, find the variance of

$$X = \int_0^1 \sqrt{t} e^{\frac{W_t^2}{8}} dW_t.$$

Answer: We will use Theorems 6.1 and 6.2 to solve this problem. To do so, we first check that the integrand $\sqrt{t} e^{\frac{W_t^2}{8}}$ is progressively measurable and square integrable in $[0, 1]$.

The process $\sqrt{t} e^{\frac{W_t^2}{8}}$ is progressively measurable because it is adapted and continuous.

Furthermore, since $W_t \sim N(0,t)$, the probability density function of W_t is $\frac{1}{\sqrt{2\pi t}} e^{-\frac{x^2}{2t}}$, and therefore

$$\begin{aligned} E\left[e^{\frac{W_t^2}{4}}\right] &= \frac{1}{\sqrt{2\pi t}} \int_{-\infty}^{\infty} e^{\frac{x^2}{4}} e^{-\frac{x^2}{2t}}\, dx \\ &= \frac{1}{\sqrt{2\pi t}} \int_{-\infty}^{\infty} e^{\left(\frac{1}{4}-\frac{1}{2t}\right)x^2}\, dx \\ &= \frac{1}{\sqrt{2\pi t}} \int_{-\infty}^{\infty} e^{\frac{1}{2}\frac{2-t}{2t}x^2}\, dx \\ &= \frac{1}{\sqrt{2\pi t}} \sqrt{2\pi \cdot \frac{2t}{2-t}} \qquad (6.141) \\ &= \sqrt{\frac{2}{2-t}}, \qquad (6.142) \end{aligned}$$

where, for (6.141), we used the identity

$$\int_{-\infty}^{\infty} e^{-\frac{Ax^2}{2}}\, dx = \sqrt{\frac{2\pi}{A}},$$

for any positive constant $A > 0$. From (6.142), it follows that

$$\begin{aligned} \int_0^1 E\left[te^{\frac{W_t^2}{4}}\right] dt &= \int_0^1 t\, E\left[e^{\frac{W_t^2}{4}}\right] dt \\ &= \int_0^1 t \cdot \sqrt{\frac{2}{2-t}} dt \\ &= -\frac{2\sqrt{2}}{3} \left((t+4)\sqrt{2-t}\right)\Big|_0^1 \\ &= \frac{2}{3}\left(8 - 5\sqrt{2}\right). \qquad (6.143) \end{aligned}$$

Thus, $\int_0^1 E\left[te^{\frac{W_t^2}{4}}\right] dt < \infty$, and we conclude that both Theorems 6.1 and 6.2 can be applied here.

From (6.136), it follows that

$$E[X] = E\left[\int_0^1 \sqrt{t} e^{\frac{W_t^2}{8}}\, dW_t\right] = 0. \qquad (6.144)$$

From (6.137), and using (6.143), we obtain that

$$\begin{aligned} E[X^2] &= E\left[\left(\int_0^1 \sqrt{t}e^{\frac{W_t^2}{8}}\,dW_t\right)^2\right] \\ &= \int_0^1 E\left[\left(\sqrt{t}e^{\frac{W_t^2}{8}}\right)^2\right] dt \\ &= \int_0^1 E\left[te^{\frac{W_t^2}{4}}\right] dt \\ &= \frac{2}{3}(8 - 5\sqrt{2}). \end{aligned} \qquad (6.145)$$

From (6.144) and (6.145), we conclude that

$$\mathrm{var}(X) = E[X^2] - (E[X])^2 = \frac{2}{3}\left(8 - 5\sqrt{2}\right). \quad \square$$

Question 18. If W_t is a Wiener process, what is $E\left[e^{W_t}\right]$?

Answer:

Solution 1: If W_t is a Wiener process and $Y = e^{W_t}$, then

$$\ln(Y) = W_t \cong \sqrt{t}Z, \qquad (6.146)$$

where Z is the standard normal random variable, and therefore Y is a lognormal random variable. Recall that the expected value of a lognormal random variable given by $\ln(V) = \mu + \sigma Z$ is

$$E[V] = e^{\mu + \frac{\sigma^2}{2}}. \qquad (6.147)$$

From (6.146), and using (6.147) with $\mu = 0$ and $\sigma = \sqrt{t}$, we conclude that

$$E\left[e^{W_t}\right] = E[Y] = e^{\frac{t}{2}}.$$

Solution 2: Since $W_t \cong \sqrt{t}Z$, it follows that

$$
\begin{aligned}
E\left[e^{W_t}\right] &= E\left[e^{\sqrt{t}Z}\right] = \frac{1}{\sqrt{2\pi}} \int_{-\infty}^{\infty} e^{\sqrt{t}x} e^{-\frac{x^2}{2}}\, dx \\
&= \frac{1}{\sqrt{2\pi}} \int_{-\infty}^{\infty} e^{\sqrt{t}x - \frac{x^2}{2}}\, dx. \qquad (6.148)
\end{aligned}
$$

By completing the square, we find that

$$
\sqrt{t}x - \frac{x^2}{2} = -\frac{1}{2}\left(x - \sqrt{t}\right)^2 + \frac{t}{2},
$$

and therefore

$$
e^{\sqrt{t}x - \frac{x^2}{2}} = e^{\frac{t}{2}} e^{-\frac{1}{2}\left(x-\sqrt{t}\right)^2}. \qquad (6.149)
$$

From (6.148) and (6.149), it follows that

$$
\begin{aligned}
E\left[e^{W_t}\right] &= e^{\frac{t}{2}} \frac{1}{\sqrt{2\pi}} \int_{-\infty}^{\infty} e^{-\frac{1}{2}\left(x-\sqrt{t}\right)^2}\, dx \\
&= e^{\frac{t}{2}} \frac{1}{\sqrt{2\pi}} \int_{-\infty}^{\infty} e^{-\frac{y^2}{2}}\, dy \qquad (6.150) \\
&= e^{\frac{t}{2}}, \qquad (6.151)
\end{aligned}
$$

where we used the substitution $y = x - \sqrt{t}$ for (6.150), and, for (6.151), we used the fact that

$$
\frac{1}{\sqrt{2\pi}} \int_{-\infty}^{\infty} e^{-\frac{y^2}{2}}\, dy = 1,
$$

since $\frac{1}{\sqrt{2\pi}} e^{-\frac{y^2}{2}}$ is the probability density function of the standard normal distribution. □

Question 19. If W_t is a Wiener process, find the variance of

$$
\int_0^t s\, dW_s.
$$

Answer: Recall that, for any deterministic square integrable function $f : [a,b] \to \mathbb{R}$, the stochastic integral $\int_a^b f(s)dW_s$ is normally distributed with mean 0 and variance equal to the square of the L^2 norm of f, i.e.,

$$\int_a^b f(s)dW_s \sim N\left(0, \int_a^b f^2(s)ds\right).$$

Then,

$$\int_0^t s\,dW_s \;\sim\; N\left(0, \int_0^t s^2 ds\right) \;=\; N\left(0, \frac{t^3}{3}\right),$$

and we conclude that

$$\text{var}\left(\int_0^t sdW_s\right) = \frac{t^3}{3}. \quad \square$$

Question 20. Let W_t be a Wiener process, and let

$$X_t \;=\; \int_0^t W_\tau d\tau. \tag{6.152}$$

What is the distribution of X_t? Is X_t a martingale?

Answer: A solution to this question was given in Chapter 2 using integration by parts; we include a different solution here.

Note that X_t is not a martingale because, if we rewrite (6.152) in differential form as

$$dX_t \;=\; W_t dt \;=\; W_t dt + 0\,dW_t,$$

we can think of X_t as being a diffusion process with only the drift part W_t.

Recall that the integral of a one-parameter family of Gaussian random variables remains Gaussian. Since W_t is a Gaussian family, X_t is normally distributed. Furthermore,

$$E[X_t] = \int_0^t E[W_\tau]d\tau = 0.$$

Therefore,

$$\mathrm{var}(X_t) = E[X_t^2] - (E[X_t])^2 = E[X_t^2]. \tag{6.153}$$

Note that

$$X_t^2 = \left(\int_0^t W_s ds\right)^2 = \int_0^t \int_0^t W_s W_u ds du, \tag{6.154}$$

and recall that

$$E[W_s W_u] = \min\{s, u\}, \ \forall\, s, u > 0; \tag{6.155}$$

see (6.125).

From (6.153–6.155), we obtain that

$$\begin{aligned}
\mathrm{var}(X_t) &= E[X_t^2] = \int_0^t \int_0^t E[W_s W_u] ds du \\
&= \int_0^t \int_0^t \min\{s, u\} ds du \\
&= \int_0^t \left(\int_0^u s ds + \int_u^t u ds\right) du \\
&= \int_0^t \left(\frac{u^2}{2} + u(t-u)\right) du \\
&= \int_0^t \left(ut - \frac{u^2}{2}\right) du \\
&= \frac{t^3}{2} - \frac{t^3}{6} \\
&= \frac{t^3}{3}.
\end{aligned}$$

We conclude that X_t is normally distributed with mean 0 and variance $\frac{t^3}{3}$, i.e.,

$$X_t \sim N\left(0, \frac{t^3}{3}\right). \quad \square$$

Question 21. What is an Itô process?

Answer: An Itô process is a generic term referring to a stochastic process X_t determined by the solution of a stochastic differential equation (SDE) of the form

$$dX_t = a(X_t, t)dt + b(X_t, t)dW_t, \qquad (6.156)$$

where W_t is a Wiener process. The coefficient $a(x,t)$ of the dt term is the drift of X_t; the coefficient $b(x,t)$ of the dW_t term is the diffusion of X_t. The SDE (6.156) is, by definition, the shorthand notation for the stochastic integral equation

$$X_t = X_0 + \int_0^t a(X_s, s)\,ds + \int_0^t b(X_s, s)\,dW_s.$$

We note that a sufficient condition for the existence and uniqueness of the (strong) solution to an SDE is for the drift and the diffusion coefficients $a(x,t)$ and $b(x,t)$ to be locally Lipschitz functions of at most linear growth in x. □

Question 22. What is Itô's lemma?

Answer: Itô's lemma, also known as Itô's formula, states that, if X_t is an Itô process satisfying the SDE

$$dX_t = a(X_t, t)dt + b(X_t, t)dW_t,$$

then for any function $f(x,t)$ with continuous second order partial derivative in x and continuous first order partial derivative in t, the process $f(X_t, t)$ is also an Itô process, driven by the same Wiener process W_t, and the drift and diffusion parts of $f(X_t, t)$ are determined according to the Taylor expansion of $f(x,t)$ to first order for the t part and up to second order for the x part:

$$\begin{aligned} df(X_t,t) &= \frac{\partial f}{\partial t}\,dt + \frac{\partial f}{\partial x}\,dX_t + \frac{1}{2}\frac{\partial^2 f}{\partial x^2}\,(dX_t)^2 \\ &= \frac{\partial f}{\partial t}\,dt + \frac{\partial f}{\partial x}\,[a(X_t,t)dt + b(X_t,t)dW_t] \\ &\quad + \frac{1}{2}\frac{\partial^2 f}{\partial x^2}\,[a(X_t,t)dt + b(X_t,t)dW_t]^2 \\ &= \left[\frac{\partial f}{\partial t} + a(X_t,t)\frac{\partial f}{\partial x} + \frac{b^2(X_t,t)}{2}\frac{\partial^2 f}{\partial x^2}\right] dt \\ &\quad + b(X_t,t)\frac{\partial f}{\partial x}\,dW_t. \end{aligned} \tag{6.157}$$

Note that for (6.157) we used the fact that

$$\begin{aligned} (dX_t)^2 &= [a(X_t,t)dt + b(X_t,t)dW_t]^2 \\ &= b^2(X_t,t)dt \end{aligned}$$

since $(dW_t)^2 = dt$ (see Problem 6.6 in this section), $(dt)^2 = 0$, and $dW_t\,dt = 0$. □

Question 23. If W_t is a Wiener process, is the process $X_t = W_t^2$ a martingale?

Answer: Recall that a stochastic process M_t defined in a filtered probability space $(\Omega, \mathcal{F}_t, P)$ is a martingale if and only if
(i) M_t is adapted, i.e., M_t is $\mathcal{F}_t$-measurable for all t;
(ii) M_t is integrable for all t, i.e., $E[M_t] < \infty$ for all t;
(iii) $E[|M_t||\mathcal{F}_s] = M_s$ almost surely for all $s < t$.

We check whether the process X_t satisfies the conditions (i), (ii), and (iii) above.

Since any continuous function of the Wiener process W_t is adapted, the process X_t is adapted, and therefore condition (i) is satisfied.

Also, $E[X_t] = E[W_t^2] = t < \infty$, since W_t is a normal random variable of mean 0 and variance t, and therefore

$E[W_t^2] = \text{var}[W_t] = t$. Thus, the random variable X_t is integrable for every t, and X_t satisfies condition (ii).

To check whether X_t satisfies condition (iii), note that, for every $s < t$,

$$\begin{aligned} E\left[X_t|\mathcal{F}_s\right] &= E\left[W_t^2|\mathcal{F}_s\right] \\ &= E\left[(W_t - W_s + W_s)^2|\mathcal{F}_s\right] \\ &= E\left[(W_t - W_s)^2|\mathcal{F}_s\right] + 2E\left[(W_t - W_s)W_s|\mathcal{F}_s\right] \\ &\quad + E\left[W_s^2|\mathcal{F}_s\right]. \end{aligned} \tag{6.158}$$

Since the Wiener process W_t has independent increments, i.e, $W_t - W_s$ is independent of $\mathcal{F}_s$, and stationary increment, i.e., $W_t - W_s \sim W_{t-s}$. Moreover, $W_t - W_s$ is a normal random variable of mean 0 and variance $t-s$, i.e., $E[W_{t-s}] = 0$ and $\text{var}(W_{t-s}) = E[W_{t-s}^2] = t - s$. Then,

$$E\left[(W_t - W_s)^2|\mathcal{F}_s\right] = E[W_{t-s}^2] = t - s, \tag{6.159}$$

and

$$\begin{aligned} E\left[(W_t - W_s)W_s|\mathcal{F}_s\right] &= W_s E\left[W_t - W_s|\mathcal{F}_s\right] \\ &= W_s E[W_{t-s}] \\ &= 0. \end{aligned} \tag{6.160}$$

Since W_s is $\mathcal{F}_s$-measurable, it follows that

$$E[W_s^2|\mathcal{F}_s] = W_s^2. \tag{6.161}$$

From (6.158–6.161), we find that

$$\begin{aligned} & E\left[X_t|\mathcal{F}_s\right] \\ = \; & E\left[(W_t - W_s)^2|\mathcal{F}_s\right] + 2E\left[(W_t - W_s)W_s|\mathcal{F}_s\right] \\ & + E\left[W_s^2|\mathcal{F}_s\right] \\ = \; & t - s + W_s^2. \end{aligned}$$

Thus,

$$E\left[X_t|\mathcal{F}_s\right] = t - s + W_s^2 \neq W_s^2 = X_s,$$

and therefore the process X_t does not satisfy condition (iii).

We conclude that X_t is not a martingale. □

Question 24. If W_t is a Wiener process, is the process

$$N_t = W_t^3 - 3tW_t$$

a martingale?

Answer:

Solution 1: A stochastic process M_t defined in a filtered probability space $(\Omega, \mathcal{F}_t, P)$ is called a martingale if and only if

(i) M_t is adapted, i.e., M_t is $\mathcal{F}_t$-measurable for all t;
(ii) M_t is integrable for all t, i.e., $E[|M_t|] < \infty$ for all t;
(iii) $E[M_t|\mathcal{F}_s] = M_s$ almost surely for all $s < t$.

We check whether the process N_t satisfies the conditions (i), (ii), and (iii) above.

Since any continuous function of the Wiener process W_t is adapted, the process N_t is adapted, and therefore condition (i) is satisfied.

Also,

$$E[N_t] = E[W_t^3] - 3tE[W_t] = 0 < \infty,$$

since $W_t \sim N(0,t)$, therefore $E[W_t] = E[W_t^3] = 0$. Thus, the random variable N_t is integrable for every t, and N_t satisfies condition (ii).

To check whether N_t satisfies condition (iii), note that, for every $s < t$,

$$\begin{aligned}
& E\left[N_t|\mathcal{F}_s\right] \\
= \ & E\left[W_t^3 - 3tW_t|\mathcal{F}_s\right] \\
= \ & E\left[(W_t - W_s + W_s)^3|\mathcal{F}_s\right] - 3tE\left[W_t|\mathcal{F}_s\right] \\
= \ & E\left[(W_t - W_s)^3|\mathcal{F}_s\right] + 3E\left[(W_t - W_s)^2 W_s|\mathcal{F}_s\right] \\
& + 3E\left[(W_t - W_s)W_s^2|\mathcal{F}_s\right] + E\left[W_s^3|\mathcal{F}_s\right] \\
& - 3tE\left[W_t|\mathcal{F}_s\right].
\end{aligned}$$

Recall that the Wiener process W_t has independent increments, i.e, $W_t - W_s$ is independent of $\mathcal{F}_s$, and stationary increment, i.e., $W_t - W_s \sim W_{t-s}$, and the fact that W_s is $\mathcal{F}_s$-measurable. Then,

$$\begin{aligned} E\left[(W_t - W_s)^3|\mathcal{F}_s\right] &= E[W_{t-s}^3] \\ &= 0; \qquad (6.162) \\ E\left[(W_t - W_s)^2 W_s|\mathcal{F}_s\right] &= W_s\, E[W_{t-s}^2] \\ &= (t-s)W_s; \qquad (6.163) \\ E\left[(W_t - W_s)W_s^2|\mathcal{F}_s\right] &= W_s^2\, E[W_{t-s}] \\ &= 0; \qquad (6.164) \\ E\left[W_s^3|\mathcal{F}_s\right] &= W_s^3, \qquad (6.165) \end{aligned}$$

since $W_{t-s} \sim N(0, t-s)$, and therefore

$$E[W_{t-s}] = E[W_{t-s}^3] = 0;$$

$$E[W_{t-s}^2] = \mathrm{var}[W_{t-s}] = t - s.$$

Moreover, since W_t is a martingale, it follows that

$$E[W_t|\mathcal{F}_s] = W_s. \qquad (6.166)$$

From (6.162–6.166), we find that

$$\begin{aligned} & E\left[N_t|\mathcal{F}_s\right] \\ = \;& E\left[(W_t - W_s)^3|\mathcal{F}_s\right] + 3E\left[(W_t - W_s)^2 W_s|\mathcal{F}_s\right] \\ & + 3E\left[(W_t - W_s)W_s^2|\mathcal{F}_s\right] + E\left[W_s^3|\mathcal{F}_s\right] \\ & - 3tE\left[W_t|\mathcal{F}_s\right] \\ = \;& 0 + 3(t-s)W_s + 0 + W_s^3 - 3tW_s \\ = \;& W_s^3 - 3sW_s \\ = \;& N_s. \end{aligned}$$

Thus, the process N_t satisfies condition (iii), and we conclude that N_t is a martingale.

Solution 2: By applying Itô's formula to N_t, we obtain that

$$\begin{aligned} dN_t &= d(W_t^3 - 3tW_t) \\ &= 3W_t^2 dW_t + \frac{1}{2} \cdot 6W_t dt - 3(W_t dt + t dW_t) \\ &= 3(W_t^2 - t) dW_t. \end{aligned}$$

The process N_t has zero drift and can be written as a stochastic integral as

$$N_t = \int_0^t 3(W_s^2 - s) dW_s.$$

Moreover,

$$\begin{aligned} & \int_0^t E\left[\left(3(W_s^2 - s)\right)^2 \right] ds \\ &= 9 \left[\int_0^t \left(E[W_s^4] - 2sE[W_s^2] + s^2 \right) ds \right] \\ &= 9 \left[\int_0^t \left(3s^2 - 2s^2 + s^2 \right) ds \right] \qquad (6.167) \\ &= 6t^3 < \infty; \end{aligned}$$

for (6.167), we used the fact that $W_s = \sqrt{s}Z$, where Z is a standard normal random variable, and therefore

$$\begin{aligned} E[W_s^2] &= E\left[\left(\sqrt{s}Z \right)^2 \right] = sE[Z^2] = s; \\ E[W_s^4] &= E\left[\left(\sqrt{s}Z \right)^4 \right] = s^2 E[Z^4] = 3s^2. \end{aligned}$$

Therefore, the process $3(W_s^2 - s)$ is square integrable for $s \in [0, t]$, and, from Theorem 6.1, we conclude that N_t is a martingale.

Question 25. What is Girsanov's theorem?

Answer: Girsanov's theorem is providing a way to change the drift of a Wiener process by defining a new probability measure via a Radon-Nikodym derivative. More precisely, let $(\Omega, \mathcal{F}_t, P)$, for $0 \leq t \leq T$, be a filtered probability space, and let W_t be a Wiener process in the probability measure P. Let h_t be a progressively measurable stochastic process such that the stochastic exponential

$$\mathcal{E}_t(h) = \exp\left(\int_0^t h_s dW_s - \frac{1}{2}\int_0^t h_s^2 ds\right)$$

is a martingale in the probability measure P. Define a new probability measure $\widetilde{P}$ over Ω given by the Radon-Nikodym derivative as follows:

$$\frac{d\widetilde{P}}{dP} = \exp\left(\int_0^T h_t dW_t - \frac{1}{2}\int_0^T h_t^2 dt\right). \tag{6.168}$$

Then, W_t is a Wiener process with drift h under the new probability measure $\widetilde{P}$. Equivalently, if we define $\widetilde{W}_t$ by $\widetilde{W}_t = W_t - h_t$, then $\widetilde{W}_t$ is a Wiener process in the $\widetilde{P}$-measure. More generally, if X_t is the diffusion process satisfying the SDE

$$dX_t = a(X_t, t)dW_t + b(X_t, t)dt$$

under the probability measure P, then under the probability measure $\widetilde{P}$ defined in (6.168), X_t satisfies the following SDE

$$\begin{aligned}
dX_t &= a(X_t, t)dW_t + b(X_t, t)dt \\
&= a(X_t, t)\left(d\widetilde{W}_t + h_t dt\right) + b(X_t, t)dt \\
&= a(X_t, t)d\widetilde{W}_t + \left(b(X_t, t) + h_t a(X_t, t)\right)dt.
\end{aligned}$$

In other words, in the P-measure, X_t has drift part b and diffusion part a; whereas in the $\widetilde{P}$-measure, the diffusion part stays the same while the drift part becomes $b + h\,a$.

In particular, if we choose h to be $-\frac{b}{a}$, then X_t becomes driftless in the $\widetilde{P}$-measure provided that the stochastic exponential $\mathcal{E}_t(h)$ is a martingale.

We note that the martingality[12] of the stochastic exponential $\mathcal{E}_t(h)$ guarantees that the new measure $\widetilde{P}$ defined in (6.168) is a probability measure, i.e., $\int_\Omega d\widetilde{P} = 1$, since

$$\begin{aligned}
\int_\Omega d\widetilde{P} &= \int_\Omega \frac{d\widetilde{P}}{dP} dP \\
&= E\left[\frac{d\widetilde{P}}{dP}\right] \\
&= E\left[\mathcal{E}_T(h)\right] \\
&= \mathcal{E}_0(h) \\
&= 1. \quad \square
\end{aligned}$$

Question 26. What is the martingale representation theorem, and how is it related to option pricing and hedging?

Answer: Let W_t be a Wiener process defined on the filtered probability space $(\Omega, \mathcal{F}_t, P)$, where the filtration $\{\mathcal{F}_t\}$ is generated by the Wiener process W_t. Let M_t be a martingale with respect to $\{\mathcal{F}_t\}$ such that M_t is square integrable for every t, i.e., $E[M_t^2] < \infty$, for all t. The martingale representation theorem asserts that, for any such a martingale M_t, there exists an $\mathcal{F}_t$-adapted square integrable process θ_t such that M_t has the stochastic integral representation

$$M_t = E[M_0] + \int_0^t \theta_s dW_s$$

almost surely. Note that such martingales have to be continuous.

---

[12] A sufficient condition which ensures the martingality of $\mathcal{E}_t(h)$ is the Novikov's condition $E\left[e^{\frac{1}{2}\int_0^T h_t^2 dt}\right] < \infty$.

The relationship between the martingale representation theorem and option pricing and hedging is as follows. For simplicity, assume that the risk free rate is zero. Assume that the price process S_t of the underlying asset follows the diffusion process determined by the SDE

$$dS_t = \sigma_t S_t dW_t$$

under risk neutral probability P, where the driving Wiener process W_t is defined on the filtered probability space $(\Omega, \mathcal{F}_t, P)$ with the filtration $\{\mathcal{F}_t\}$ is generated by W_t. Consider a derivative whose payoff function (possibly path dependent) is φ_T at maturity T. The Fundamental Theorem of Asset Pricing asserts that, because the risk free rate is assumed zero, the value process V_t of the derivative security is a martingale under risk neutral probability. In fact,

$$V_t = E[\varphi_T | \mathcal{F}_t].$$

Moreover, if the payoff function φ_T is square integrable, i.e., $E[\varphi_T^2] < \infty$, then V_t is a square integrable martingale. Therefore, by applying the martingale representation theorem, we find that there exists a $\mathcal{F}_t$-adapted process θ_t such that

$$\begin{aligned}\varphi_T &= V_T = E[V_0] + \int_0^T \theta_t dW_s \\ &= E[V_0] + \int_0^T \frac{\theta_t}{\sigma_t S_t} dS_t.\end{aligned}$$

Note that the quantity $\frac{\theta_t}{\sigma_t S_t}$ indicates the amount of shares to hold in order to dynamically hedge the position of the derivative with payoff function φ_T at maturity T. In particular, if φ_T is the payoff of a call option, then $\frac{\theta_t}{\sigma_t S_t}$ corresponds to the delta of the call. $\square$

Question 27. Solve

$$dY_t = Y_t dW_t, \tag{6.169}$$

where W_t is a Wiener process.

Answer:
Solution 1: Note that (6.169) is a particular case for $\mu = 0$ and $\sigma = 1$ of the stochastic differential equation

$$dS_t \;=\; \mu S_t dt \;+\; \sigma S_t dW_t, \tag{6.170}$$

which is the model for the evolution of an asset in the Black–Scholes framework. The solution of (6.170) has the distribution

$$S_t \;=\; S_0 \exp\left(\left(\mu - \frac{\sigma^2}{2}\right)t + \sigma\sqrt{t}Z\right), \tag{6.171}$$

with $t > 0$, where Z is the standard normal variable. For $\mu = 0$ and $\sigma = 1$ in (6.171), we obtain that the solution to (6.169) is

$$Y_t \;=\; Y_0 \exp\left(-\frac{t}{2} + \sqrt{t}Z\right). \tag{6.172}$$

Solution 2: From Itô's lemma it follows that, if Y_t is a stochastic process satisfying $dY_t = Y_t dW_t$ and $f(y)$ is a function with continuous second order derivative, then

$$d(f(Y_t)) \;=\; \frac{1}{2}f''(Y_t)Y_t^2 dt \;+\; f'(Y_t)Y_t dW_t. \tag{6.173}$$

For $f(y) = \ln(y)$, we obtain from (6.173) that

$$d(\ln(Y_t)) \;=\; -\frac{1}{2}dt \;+\; dW_t. \tag{6.174}$$

By integrating (6.174) between 0 and t, we find that

$$\ln(Y_t) - \ln(Y_0) \;=\; -\frac{t}{2} + W_t - W_0 \;=\; -\frac{t}{2} + W_t,$$

since $W_0 = 0$, and therefore

$$Y_t \;=\; Y_0 \exp\left(-\frac{t}{2} + W_t\right).$$

Since W_t is a Wiener process, W_t is a normal random variable of mean 0 and variance t, i.e., $W_t = \sqrt{t}Z$, where Z is the standard normal variable. Thus, Y_t has the distribution

$$Y_t = Y_0 \exp\left(-\frac{t}{2} + \sqrt{t}Z\right),$$

which is the same as (6.172). □

Question 28. Solve the following SDEs:

(i) $dY_t = \mu Y_t dt + \sigma Y_t dW_t$;

(ii) $dX_t = \mu dt + (aX_t + b)dW_t$.

Answer:
(i) Recall from Itô's formula that, if Y_t satisfies the SDE

$$dY_t = \mu Y_t dt + \sigma Y_t dW_t,$$

and if $f(y)$ is a function with continuous second order derivative, then

$$\begin{aligned} df(Y_t) &= f'(Y_t)dY_t + \frac{1}{2}f''(Y_t)\sigma^2 Y_t^2 dt \\ &= \left[\mu Y_t f'(Y_t) + \frac{\sigma^2}{2}Y_t^2 f''(Y_t)\right] dt \\ &\quad + \sigma Y_t f'(Y_t)dW_t. \end{aligned} \tag{6.175}$$

For $f(y) = \ln(y)$, we obtain from (6.175) that

$$d\ln(Y_t) = \left(\mu - \frac{\sigma^2}{2}\right) dt + \sigma dW_t. \tag{6.176}$$

By integrating (6.176) from 0 to t, we obtain that

$$\ln(Y_t) - \ln(Y_0) = \left(\mu - \frac{\sigma^2}{2}\right) t + \sigma W_t, \tag{6.177}$$

and by solving (6.177) for Y_t, we conclude that

$$Y_t = Y_0 e^{\left(\mu - \frac{\sigma^2}{2}\right)t + \sigma W_t}.$$

(ii) We look for a solution of

$$dX_t = \mu dt + (aX_t + b)dW_t \tag{6.178}$$

of the form

$$X_t = U_t V_t, \tag{6.179}$$

where the process U_t is defined by the solution to the SDE

$$dU_t = aU_t dW_t, \quad \text{with } U_0 = 1, \tag{6.180}$$

and the process V_t is the solution to the SDE

$$dV_t = \alpha_t dt + \beta_t dW_t, \quad \text{with } V_0 = X_0, \tag{6.181}$$

where the coefficients α_t and β_t are to be determined.

Recall from (i) that the solution to

$$dY_t = \mu Y_t dt + \sigma Y_t dW_t$$

is

$$Y_t = Y_0\, e^{\left(\mu - \frac{\sigma^2}{2}\right)t + \sigma W_t}. \tag{6.182}$$

By letting $\mu = 0$ and $\sigma = a$ in (6.182), we find that the solution to (6.180) is

$$U_t = e^{aW_t - \frac{a^2}{2}t}. \tag{6.183}$$

Note that

$$\begin{aligned} dU_t\, dV_t &= (aU_t dW_t)(\alpha_t dt + \beta_t dW_t) \\ &= a\beta_t U_t dt, \end{aligned} \tag{6.184}$$

since $(dW_t)^2 = dt$ and $dW_t dt = 0$ (because this term has order $(dt)^{3/2}$). By applying Itô's product rule to $X_t = U_t V_t$ and using (6.184), we obtain that

$$\begin{aligned} dX_t &= d(U_t V_t) = U_t dV_t + V_t dU_t + dU_t\, dV_t \\ &= U_t(\alpha_t dt + \beta_t dW_t) + V_t(aU_t dW_t) \\ &\quad + a\beta_t U_t dt \\ &= (a\beta_t + \alpha_t)U_t dt + (aU_t V_t + \beta_t U_t)dW_t \\ &= (a\beta_t + \alpha_t)U_t dt + (aX_t + \beta_t U_t)dW_t \quad (6.185) \end{aligned}$$

Then, $X_t = U_t V_t$ is a solution to (6.178) if and only if the coefficients of the dW_t terms and of the dt terms in (6.178) and (6.185) are equal.

From the dW_t terms, we obtain that

$$\begin{aligned} & aX_t + b = aX_t + \beta_t U_t \\ \Longleftrightarrow\quad & \beta_t U_t = b \\ \Longleftrightarrow\quad & \beta_t = bU_t^{-1} = b\, e^{-aW_t + \frac{a^2}{2}t}. \end{aligned} \tag{6.186}$$

From the dt terms, we obtain that

$$\begin{aligned} & (a\beta_t + \alpha_t)U_t = \mu \\ \Longleftrightarrow\quad & a\beta_t + \alpha_t = \mu U_t^{-1} \\ \Longleftrightarrow\quad & \alpha_t = \mu\, U_t^{-1} - a\beta_t \\ \Longleftrightarrow\quad & \alpha_t = \mu\, e^{-aW_t + \frac{a^2}{2}t} - abe^{-aW_t + \frac{a^2}{2}t} \\ \Longleftrightarrow\quad & \alpha_t = (\mu - ab)\, e^{-aW_t + \frac{a^2}{2}t}. \end{aligned} \tag{6.187}$$

Note that the solution V_t to (6.181) is given by

$$V_t = X_0 + \int_0^t \alpha_s ds + \int_0^t \beta_s dW_s. \tag{6.188}$$

From (6.179), (6.183), and (6.188), we find that the solution X_t to (6.178) is

$$X_t = e^{aW_t - \frac{a^2}{2}t}\left(X_0 + \int_0^t \alpha_s ds + \int_0^t \beta_s dW_s\right),$$

and, using (6.186) and (6.187), we obtain that

$$\begin{aligned} X_t &= e^{aW_t - \frac{a^2}{2}t}\left[X_0 + \int_0^t (\mu - ab)\, e^{-aW_s + \frac{a^2}{2}s} ds \right. \\ &\quad \left. + \int_0^t be^{-W_s + \frac{a^2}{2}s} dW_s \right] \\ &= X_0\, e^{aW_t - \frac{a^2}{2}t} \\ &\quad + (\mu - ab)\int_0^t e^{a(W_t - W_s) - \frac{a^2}{2}(t-s)} ds \\ &\quad + b\int_0^t e^{a(W_t - W_s) - \frac{a^2}{2}(t-s)} dW_s. \quad \square \end{aligned}$$

Question 29. What is the Heston model?

Answer: In Heston's stochastic volatility model, it is assumed that the price of the underlying asset satisfies the same SDE as in the lognormal model, i.e.,

$$dS_t = \mu S_t dt + \sqrt{v_t} S_t dW_t,$$

whereas the instantaneous variance v_t itself follows a mean reverting CIR (Cox-Ingersoll-Ross) process

$$dv_t = \lambda(v_t - m)dt + \eta\sqrt{v_t} dZ_t,$$

where $\lambda > 0$ and $m > 0$ are positive constants. The driving Wiener processes W_t and Z_t are correlated with constant correlation ρ, i.e.,

$$\mathrm{corr}(dW_t, dZ_t) = \rho dt.$$

The Heston model is a benchmark, and is commonly used in derivative pricing because it has following features:

- it takes into account the leverage effect, namely, the driving Wiener processes W_t and Z_t are correlated; empirically, the correlation ρ is negative and that is why the word "leverage";

- it has a quasi closed form (up to an inverse Fourier transform) solution for the prices of European options, which make the calibration more tractable;
- the variance process is mean reverting with rate of reversion λ and long term mean m.

Note that, if the parameters of the volatility process are in the regime $2\lambda m < \eta^2$, then zero is an attainable boundary for the volatility process. Practitioners usually assume that the boundary behavior at zero is either absorption, i.e., the process is stuck at zero once it hits zero, or reflection, i.e, the process bounces back right after it hits zero. On the other hand, the other boundary infinity is an unattainable boundary. $\square$

6.7 Brainteasers.

Question 1. A flea is going between two points which are 100 inches apart by jumping (always in the same direction) either one inch or two inches at a time. How many different paths can the flea travel by?

Answer: Let a_n denote the number of different paths of the flea that covers the distance of n inches, jumping either one inch or two inches at a time. We want to find a_{100}.

Since the flea can jump either one inch or two inches at a time, it could have made the last jump either from the end of the $(n-1)$st inch or from the end of the $(n-2)$nd inch. Hence, the total number of ways the flea can cover the distance of n inches, jumping either one inch or two inches at a time, is the sum of the number of ways the flea can cover the distance of $n-1$ inches, jumping either one inch or two inches at a time, and the number of ways the flea can cover the distance of $n-2$ inches, jumping either one inch or two inches at a time.

In other words, for $n > 2$, we have

$$a_n = a_{n-1} + a_{n-2}. \tag{6.189}$$

Note that $a_1 = 1$, since the flea can cover 1 inch in only one way, jumping one inch once; while $a_2 = 2$, since the flea can cover 2 inches in two ways, either jumping one inch twice, or jumping two inches once.

Note that (6.189) is the recurrence relation for the Fibonacci sequence. Let

$$\phi_1 = \frac{1+\sqrt{5}}{2} \quad \text{and} \quad \phi_2 = \frac{1-\sqrt{5}}{2}$$

be the roots of the characteristic equation $x^2 - x - 1 = 0$ corresponding to (6.189).

Then,

$$a_n = C_1\phi_1^n + C_2\phi_2^n, \ \ \forall\, n \geq 1,$$

where the constants C_1 and C_2 are such that $a_1 = 1$ and $a_2 = 2$.

By solving the linear system

$$\begin{cases} C_1\phi_1 + C_2\phi_2 & = & 1 \\ C_1\phi_1^2 + C_2\phi_2^2 & = & 2, \end{cases}$$

we obtain that

$$\begin{aligned} C_1 &= \frac{1+\sqrt{5}}{2\sqrt{5}} = \frac{\phi_1}{\sqrt{5}}; \\ C_2 &= -\frac{\sqrt{5}-1}{2\sqrt{5}} = -\frac{\phi_2}{\sqrt{5}}, \end{aligned}$$

and therefore

$$a_n = \frac{1}{\sqrt{5}}\left(\phi_1^{n+1} - \phi_2^{n+1}\right).$$

Plugging $n = 100$ into the last expression gives the answer. □

Question 2. I have a bag containing three pancakes: one golden on both sides, one burnt on both sides, and one golden on one side and burnt on the other. You shake the bag, draw a pancake at random, look at one side, and notice that it is golden. What is the probability that the other side is golden?

Answer:

Solution 1: Label the pancakes 0, 1, and 2, according to the number of burnt sides. Let E_i, $i = 0, 1, 2$, denote the event that the pancake with i burnt sides is drawn from the bag. Let A denote the event that the side (of the randomly drawn pancake) we look at is golden.

The probability of the pancake drawn from the bag having no burnt sides given that one side is golden is $P(E_0|A)$. Then, from the Bayes'formula, we obtain that

$$P(E_0|A) = \frac{P(E_0 \cap A)}{P(A)} = \frac{P(E_0)P(A|E_0)}{P(A|E_0)P(E_0) + P(A|E_1)P(E_1) + P(A|E_2)P(E_2)}$$
$$= \frac{\frac{1}{3} \cdot 1}{1 \cdot \frac{1}{3} + \frac{1}{2} \cdot \frac{1}{3} + 0 \cdot \frac{1}{3}}$$
$$= \frac{2}{3}.$$

Solution 2: Out of the six possible sides that we could have seen, three are golden. Out of these, two belong to a pancake that is golden on both sides. Therefore, the probability of the other side being golden is $\frac{2}{3}$. □

Question 3. Alice and Bob are playing heads and tails, Alice tosses $n+1$ coins, Bob tosses n coins. The coins are fair. What is the probability that Alice will have strictly more heads than Bob?

Answer:
Solution 1: Alice flips the coin more often than Bob, so either she must end up with more heads or with more tails than Bob. She cannot, however, end up with more heads and more tails, because she only flips one more coin than Bob. We deduce that either Alice gets more heads or she gets more tails.

Since these events are equally likely, they both have probability $\frac{1}{2}$, and therefore the probability that Alice will have strictly more heads than Bob is $\frac{1}{2}$.

Solution 2: Suppose that Alice and Bob begin by flipping n coins each. Let p be the probability that Alice gets more heads than Bob, and let q be the probability that both Alice and Bob get an equal number of heads. Note that $2p + q = 1$ since the probability that Alice gets more heads than Bob is, by symmetry, equal to the probability that Bob gets more heads than Alice.

Alice then flips the $(n+1)$st coin. For Alice to have more heads, she either had more heads than Bob before flipping the last coin; or the same number of heads as Bob before flipping the last coin and her $(n+1)$st flip must come up heads. Hence, the probability of Alice winning is $p + \frac{1}{2}q$.

Since $2p + q = 1$, then $p + \frac{1}{2}q = \frac{1}{2}$. □

Question 4. Alice is in a restaurant trying to decide between three desserts. How can she choose one of three desserts with equal probability with the help of a fair coin? What if the coin is biased and the bias is unknown?

Answer:

Solution 1: Denote the desserts by A, B, and C. First, suppose the coin is fair. Denote heads by H and tails by T. The procedure to choose one of three desserts with equal probability is as follows: toss the coin twice; let the outcomes TH, HT, and TT correspond to choosing desserts A, B, and C, respectively; if the outcome is HH, repeat the procedure.

Note that the probability of our procedure not being repeated is $p = \frac{3}{4}$; hence, the number of times our procedure is repeated is a geometric random variable with p as its parameter. The expected number of times our procedure is repeated is $\frac{1}{p} = \frac{4}{3}$. Since each procedure involves two coin tosses, the expected number of coin tosses before Alice chooses one of three desserts with equal probability is $\frac{8}{3}$.

Now, suppose the coin is biased. One procedure to choose one of three desserts with equal probability would be as follows: toss the coin four times; denote by $THHT$, $HTTH$, and $THTH$ the outcomes corresponding to choosing desserts A, B, and C, respectively; all the other 4-toss outcomes result in repeating the procedure.

Solution 2: An alternative procedure is as follows: toss the coin three times; denote by HTT, THT, and TTH the outcomes corresponding to choosing desserts A, B, and C, respectively; all the other 3-toss outcomes result in repeating the procedure.

Using an argument similar to the case with a fair coin, one finds that the expected number of tosses of a coin with an unknown bias, before Alice chooses one of three desserts with equal probability, is $\frac{64}{3}$ and 8, respectively, for the two procedures described above. □

Question 5. What is the expected number of times you must flip a fair coin until it lands on head? What if the coin is biased and lands on head with probability p?

Answer: Denote by X be the number of times you must flip a fair coin until it lands on head. If the first coin toss is a head (which happens with probability $\frac{1}{2}$), then $X = 1$. If the first coin toss is a tail (which also happens with probability $\frac{1}{2}$), then the coin tossing process resets and the number of steps before the coin lands on head will be 1 plus the expected number of coin tosses until the coin lands heads. In other words, the expected number of coin toss $E[X]$ satisfies the equation

$$E[X] \;=\; \frac{1}{2} \;+\; \frac{1}{2}\,(1 + E[X]). \qquad (6.190)$$

Solving (6.190) for $E[X]$, we conclude that $E[X] = 2$, i.e., the expected number of times you must flip a fair coin until it lands heads is 2.

On the other hand, if the coin is biased with the probability p of landing on head, the same argument still applies. However, in this case, (6.190) reads

$$E[X] \;=\; p \;+\; (1-p)\,(1 + E[X]).$$

Again, solve for $E[X]$, we obtain $E[X] = \frac{1}{p}$. □

Question 6. What is the expected number of coin tosses of a fair coin in order to get two heads in a row? What if the coin is biased with 25% probability of getting heads?

Answer: We solve the general case of a biased coin with probability p of the coin toss resulting in heads. The outcomes of the first two tosses are as follows:

• If the first toss is tails, which happens with probability $1-p$, then the process resets and the expected number of tosses increases by 1.

• If the first toss is heads, and if the second toss is also heads, which happens with probability p^2, then two consecutive heads are obtained after two tosses.

• If the first toss is heads, and if the second toss is tails, which happens with probability $p(1-p)$, then the process resets and the expected number of tosses increases by 2.

If $E[X]$ denotes the expected number of tosses in order to get two heads in a row, we conclude that

$$E[X] = (1-p)(1+E[X]) + 2p^2 + p(1-p)(2+E[X]).$$

By solving for $E[X]$, we obtain that

$$E[X] = \frac{1+p}{p^2}. \tag{6.191}$$

For an unbiased coin, i.e., for $p = \frac{1}{2}$, we find from (6.191) that $E[X] = 6$, i.e., the expected number of coin tosses to obtain two heads in a row is 6.

For a biased coin with 25% probability of getting heads, i.e., for $p = \frac{1}{4}$, we find from (6.191) that $E[X] = 20$, i.e., the expected number of coin tosses to obtain two heads in a row in this case is 20. □

Question 7. A fair coin is tossed n times. What is the probability that no two consecutive heads appear?

Answer: The total number of sequences of heads and tails of length n is 2^n. Let a_n be the number of sequences of heads and tails of length n, such that no two consecutive heads appear. Then, the probability that no two consecutive heads appear is $\frac{a_n}{2^n}$.

Note that $a_1 = 2$ (H and T do not contain two consecutive heads) and $a_2 = 3$ (out of HH, HT, TH, and TT, only HH contains two consecutive heads). We find the closed formula for a_n by deriving a recurrence relation as follows.

A sequence of $n \geq 3$ coin tosses does not contain two consecutive heads if and only if: (i) either it begins with a tail, followed by a sequence of $n-1$ coin tosses with no two consecutive heads; (ii) or it begins with a head, followed by a tail, and followed by a sequence of $n-2$ coin tosses with no two consecutive heads. Since these two scenarios are mutually exclusive, it follows that

$$a_n = a_{n-1} + a_{n-2}, \quad \forall\, n \geq 3. \tag{6.192}$$

Note that (6.192) is the recurrence relation for the Fibonacci sequence. Let $\phi_1 = \frac{1+\sqrt{5}}{2}$ and $\phi_2 = \frac{1-\sqrt{5}}{2}$ be the roots of the characteristic equation $x^2 - x - 1 = 0$ corresponding to (6.192). Then,

$$a_n = C_1\phi_1^n + C_2\phi_2^n, \quad \forall\, n \geq 1,$$

where the constants C_1 and C_2 are such that $a_1 = 2$ and $a_2 = 3$.

By solving the linear system

$$\begin{cases} C_1\phi_1 + C_2\phi_2 = 2 \\ C_1\phi_1^2 + C_2\phi_2^2 = 3, \end{cases}$$

we obtain that

$$C_1 = \frac{3+\sqrt{5}}{2\sqrt{5}} = \frac{\phi_1^2}{\sqrt{5}};$$

$$C_2 = \frac{3-\sqrt{5}}{2\sqrt{5}} = -\frac{\phi_2^2}{\sqrt{5}}.$$

We conclude that

$$a_n = \frac{1}{\sqrt{5}}\left(\phi_1^{n+2} - \phi_2^{n+2}\right),$$

and therefore the probability that no two consecutive heads appear in n tosses of a fair coin, which is equal to $\frac{a_n}{2^n}$, is

$$\frac{1}{2^n\sqrt{5}}\left(\phi_1^{n+2} - \phi_2^{n+2}\right). \quad \square$$

Question 8. You have two identical Fabergé eggs, either of which would break if dropped from the top of a building with 100 floors. Your task is to determine the highest floor from which an egg could be dropped without breaking. What is the minimum number of drops required to achieve this? You are allowed to break both eggs in the process.

Answer: Consider the following more general problem:

Find the largest number of floors $h_e(n)$ a building could have in order to be able to determine the highest floor from which the an egg could be dropped without breaking using e eggs and n drops.

Since one drop can only determine one floor, it follows that

$$h_e(1) = 1. \tag{6.193}$$

If we have only one egg at our disposal, the only possible strategy is to try the floors one by one from bottom to top; hence,

$$h_1(n) = n.$$

When $e \geq 2$ and $n \geq 2$, the first drop cannot be from the floor higher than $h_{e-1}(n-1)+1$, since if the egg breaks, there are only $e-1$ eggs and $n-1$ drops left, and the highest floor we can still handle is $h_{e-1}(n-1)$. If the first drop does not break an egg, we can treat floor

$h_{e-1}(n-1)+2$ as the new floor 1, and reduce it to a problem with e eggs and $n-1$ drops, and therefore

$$h_e(n) \;=\; 1 + h_{e-1}(n-1) + h_e(n-1).$$

Iterating this argument, we obtain that

$$\begin{aligned} & h_e(n) \\ = \;& 1 + h_{e-1}(n-1) + h_e(n-1) \\ = \;& 2 + h_{e-1}(n-1) + h_{e-1}(n-2) + h_e(n-2) \\ = \;& \ldots \\ = \;& (n-1) + \sum_{j=1}^{n-1} h_{e-1}(j) + h_e(1) \\ = \;& n + \sum_{j=1}^{n-1} h_{e-1}(j), \end{aligned}$$

since $h_e(1) = 1$; see (6.193).

For $e = 2$, the formula above becomes

$$\begin{aligned} & h_2(n) \\ = \;& n + \sum_{j=1}^{n-1} h_1(j) = n + \sum_{j=1}^{n-1} j = n + \frac{(n-1)n}{2} \\ = \;& \frac{n(n+1)}{2}, \end{aligned} \tag{6.194}$$

where the next-to-last step follows from the summation formula $\sum_{j=1}^{k} = \frac{k(k+1)}{2}$.

Since $h_2(13) = 91 < 100 \leq 105 = h_2(14)$, the required number of drops is 14.

Note that our iterative argument also provides an algorithm: you drop the first egg from the floors

$$\begin{aligned}
14 &= 1 + h_1(13), \\
27 &= 2 + h_1(13) + h_1(12), \\
39 &= 3 + \sum_{j=11}^{13} h_1(j), \\
50 &= 4 + \sum_{j=10}^{13} h_1(j), \\
60 &= 5 + \sum_{j=9}^{13} h_1(j), \\
69 &= 6 + \sum_{j=8}^{13} h_1(j), \\
77 &= 7 + \sum_{j=7}^{13} h_1(j), \\
84 &= 8 + \sum_{j=6}^{13} h_1(j), \\
90 &= 9 + \sum_{j=5}^{13} h_1(j), \\
95 &= 10 + \sum_{j=4}^{13} h_1(j), \\
99 &= 11 + \sum_{j=3}^{13} h_1(j),
\end{aligned}$$

and 100; that is, move up by $14 = 1 + h_1(13)$ floors, then by $13 = 1 + h_1(12)$ floors, then by $12 = 1 + h_1(11)$ floors, and so on, until the first egg breaks (or does not) from the 100th floor. Calling the floor from which the first egg breaks f and the previously tested floor f', you drop the second egg from the intervening floors $f'+1, f'+2, \ldots, f-1$ in that order. $\square$

Question 9. An ant is in the corner of a $10 \times 10 \times 10$ room and wants to go to the opposite corner. What is the length of the shortest path the ant can take?

Answer: For clarity, assume that the ant is in a corner by the ceiling. Denote that corner by A, and denote by B the opposite corner to A, which is by the floor. The shortest path from A to B would require the ant to go on a straight line across a wall of the room to a side of the floor and from there on a straight line along the floor to the B. If you imagine laying down to the floor the vertical wall the ant went down from A to the floor, you have a 10×20 rectangle with A and B opposite corners in the rectangle. The shortest path for the ant to go from A to B is by following the diagonal of the rectangle, which has length $10\sqrt{5} \approx 22.36$. $\square$

Question 10. A $10 \times 10 \times 10$ cube is made of $1,000$ unit cubes. How many unit cubes can you see on the outside?

Answer: If all the outside unit cubes are removed, what remains is an $8 \times 8 \times 8$ cube, which is made of $8^3 = 512$ unit cubes. Thus, there are $1000 - 512 = 488$ outside unit cubes. $\square$

Question 11. Fox Mulder is imprisoned by aliens in a large circular field surrounded by a fence. Outside the fence is a vicious alien that can run four times as fast as Mulder, but is constrained to stay near the fence. If Mulder can contrive to get to an unguarded point on the fence, he can quickly scale the fence and escape. Can he get to a point on the fence ahead of the alien?

Answer: Let R denote the radius of the circular field, whose center we denote by C. Denote Mulder's speed by

v. The alien's speed is then $4v$. Denote Mulder's and alien's positions by M and A, respectively.

Mulder cannot just run for the fence along the straight line connecting C with the point on the fence diametrically opposite to A. Indeed, while it takes Mulder $\frac{R}{v}$ time to cover the distance R, the alien would cover the distance πR in $\frac{\pi R}{4v}$ time and the alien would catch up with Mulder, since $\frac{R}{v} > \frac{\pi R}{4v}$.

To optimize this strategy, Mulder needs to start running for the fence from a point that is closer to the fence than C. Assume that Mulder somehow managed to be at a point M that is xR away from C (where $0 < x < 1$), with M, C, and A collinear, and C between M and A. Denote by P the point on the circle diametrically opposite to A; see Figure 6.2. Then, $MC = xR$ and $MP = (1-x)R$. It takes Mulder $\frac{(1-x)R}{v}$ time to reach the fence running from M to P, while the alien needs $\frac{\pi R}{4v}$ time to reach the point P going from A to P on a semicircle. Note that

$$\frac{(1-x)R}{v} < \frac{\pi R}{4v}, \quad \text{if } x > 1 - \frac{\pi}{4}.$$

Thus, if $x > 1 - \frac{\pi}{4}$, Mulder would be able to escape the alien.

We are now ready to describe Mulder's escape strategy. He sets $x = 1 - \frac{\pi}{4} + 0.01$. Note that $x < \frac{1}{4}$, since $0.01 < \frac{\pi - 3}{4}$. Regardless of the alien's movement, Mulder first runs from C to any point on the circle of radius xR, centered at C. Then he runs around that circle, until his position M is such that M, C, and A are collinear, with C between M and A. He is able to do so, since $x < \frac{1}{4}$ and the alien's speed is only 4 times his speed.[13] Finally,

[13]Mulder is able to do so since, if $x < \frac{1}{4}$, his angular speed, $\frac{\theta v}{\pi R x}$, is larger than the angular speed $\frac{4\theta v}{\pi R}$ of the alien:

$$\frac{\theta v}{\pi R x} > \frac{4\theta v}{\pi R} \iff \frac{1}{4} > x.$$

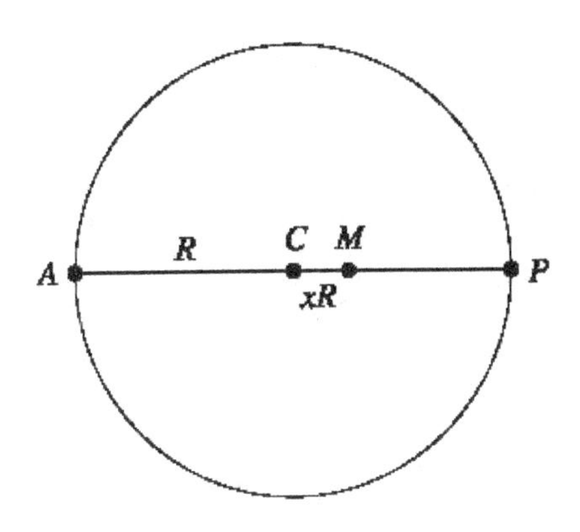

图 6.2: Mulder can reach P from M before the alien can do so from A.

Mulder runs from M to P and will reach P before alien does, as shown above, since $x > 1 - \frac{\pi}{4}$. □

Question 12. At your subway station, you notice that of the two trains running in opposite directions which are supposed to arrive with the same frequency, the train going in one direction comes first 80% of the time, while the train going in the opposite direction comes first only 20% of the time. What do you think could be happening?

Answer: One thing that could be happening is that the train that comes first 80% of the time comes in fact more frequently than the other one. However, even if both trains run with the same frequency, one train might come first 80% of the time. For example, assuming that your arrival in the station is uniformly distributed, if both trains run every ten minutes, and train A comes into the station at 1:00, 1:10, 1:20, ..., while train B comes at 1:12, 1:22,

1:32, ..., then train A will come first 80% of the time. □

Question 13. You start off with one amoeba. Every minute, this amoeba can either die, do nothing, split into two amoebas, or split into three amoebas; all these scenarios being equally likely to happen. All further amoebas behave the same way. What is the probability that the amoebas eventually die off?

Answer:
Solution 1: (Due to Yu Gan, Baruch MFE'14.) Denote by A_k the event that no amoebas are alive after k minutes. Let $p_k = P(A_k)$. Note that $p_1 = \frac{1}{4}$ and $A_k \subseteq A_{k+1}$, for all $k \geq 1$.

The probability p that the amoebas eventually die off is the probability that at some point in time, i.e., after n minutes for some n, no amoebas are alive. In other words,

$$p = P\left(\bigcup_{k=1}^{\infty} A_k\right).$$

Note that $\bigcup_{k=1}^{n} A_k = A_n$, since $A_k \subseteq A_{k+1}$ for all $k \geq 1$. Then,

$$\begin{aligned} p &= P\left(\bigcup_{k=1}^{\infty} A_k\right) = P\left(\lim_{n\to\infty} \bigcup_{k=1}^{n} A_k\right) \\ &= \lim_{n\to\infty} P\left(\bigcup_{k=1}^{n} A_k\right) = \lim_{n\to\infty} P(A_n) \\ &= \lim_{n\to\infty} p_n. \end{aligned} \tag{6.195}$$

Given the four equally probable outcomes after the first minute, i.e, the amoeba can die, remain one amoeba, split into two amoebas, or split into three amoebas, it follows that

$$p_n = \frac{1}{4} + \frac{1}{4}p_{n-1} + \frac{1}{4}p_{n-1}^2 + \frac{1}{4}p_{n-1}^3, \tag{6.196}$$

for all $n \geq 2$.

The sequence $(p_n)_{n\geq 0}$ is increasing. Recall that $A_n \subseteq A_{n+1}$, and therefore

$$p_n = P(A_n) \leq P(A_{n+1}) = p_{n+1}, \ \forall\, n \geq 1.$$

Also, we can see by induction that the sequence $(p_n)_{n\geq 0}$ is bounded from above by $\sqrt{2}-1$, since $p_1 = \frac{1}{4} < \sqrt{2}-1$, and, if we assume that $p_{n-1} < \sqrt{2}-1$ for some $n \geq 2$, then we obtain from (6.196) that

$$\begin{aligned} p_n &< \frac{1}{4} + \frac{1}{4}(\sqrt{2}-1) + \frac{1}{4}(\sqrt{2}-1)^2 + \frac{1}{4}(\sqrt{2}-1)^3 \\ &= \sqrt{2}-1. \end{aligned}$$

Thus, the sequence $(p_n)_{n\geq 0}$ is bounded from above and increasing, and therefore convergent.

Recall from (6.195) that $\lim_{n\to\infty} p_n = p$, where p is the probability that the amoebas eventually die off. Since $p_n < \sqrt{2}-1$ for all $n \geq 1$, it follows that $p \leq \sqrt{2}-1$. Moreover, from (6.196), we find that

$$p = \frac{1}{4} + \frac{1}{4}p + \frac{1}{4}p^2 + \frac{1}{4}p^3,$$

which can be written as

$$0 = p^3 + p^2 - 3p + 1 = (p-1)(p^2 + 2p - 1),$$

and has solutions 1, $\sqrt{2}-1$, and $-\sqrt{2}-1$. Since $0 < p \leq \sqrt{2}-1$, we obtain that $p = \sqrt{2}-1$.

In other words, the probability that the amoebas eventually die off is $\sqrt{2}-1$.

Solution 2: Let p be the probability that the descendants of a single amoeba will die out eventually. Then, the probability that the descendants of n amoebas will all die out eventually is p^n, since each amoeba is independent of all other amoebas.

Furthermore, the probability that the descendants of an amoeba will die out eventually is independent of time when averaged over all the possibilities.

At the beginning, the probability that the descendants of an amoeba will die out eventually is, by definition, p. After one minute, the initial amoeba turns into 0, 1, 2, or 3 amoebas, with probability of $\frac{1}{4}$ for each case. Thus, the probability that the descendants of the amoeba die out is now

$$\frac{1}{4}\left(p^0 + p^1 + p^2 + p^3\right) = \frac{1}{4}\left(1 + p + p^2 + p^3\right).$$

These two probabilities must be equal, and therefore

$$p = \frac{1}{4}\left(1 + p + p^2 + p^3\right), \tag{6.197}$$

which is the same as

$$p^3 + p^2 - 3p + 1 = (p-1)(p^2 + 2p - 1) = 0. \tag{6.198}$$

The roots of (6.198) are $p = 1$, $p = -\sqrt{2} - 1$, and $p = \sqrt{2} - 1$. The only root in the interval $(0, 1)$ is $p = \sqrt{2} - 1$, and this is the probability that the amoebas eventually die off.

A more subtle question is why $p = 1$ is not one of the possible answers to the problem in hand? The right hand side of (6.197) is the generating function $h(p)$ of amoeba's branching process. It is a well-known theorem on branching processes that if the mean number of offspring produced by a single amoeba is bigger than 1, then the smallest positive root of the equation $p = h(p)$ is the probability that amoeba's descendants will die out eventually. In our case, the mean number of offspring produced by a single amoeba is $\frac{1}{4}(0 + 1 + 2 + 3) = \frac{3}{2} > 1$, so the theorem applies. □

Question 14. Given a set X with n elements, choose two subsets A and B at random. What is the probability of A being a subset of B?

Answer: When two subsets A and B of X are chosen at random, each element of X is equally likely to end up in any of the following four sets:

$$A \setminus B, \ B \setminus A, \ A \cap B, \text{ and } X \setminus (A \cup B).$$

For A to be a subset of B, $A \setminus B$ would have to be empty; in other words, none of the n elements of X would end up in $A \setminus B$. The probability of any element of X not ending up in $A \setminus B$ is $\frac{3}{4}$.

We conclude that the probability of A being a subset of B is

$$\left(\frac{3}{4}\right)^n. \quad \square$$

Question 15. Alice writes two distinct real numbers between 0 and 1 on two sheets of paper. Bob selects one of the sheets randomly to inspect it. He then has to declare whether the number he sees is the bigger or smaller of the two.

Is there any way Bob can expect to be correct more than half the times Alice plays this game with him?

Answer: Denote the numbers Alice writes on two sheets of paper by a_1 and a_2, $0 < a_1 < a_2 < 1$. Denote the number Bob selects by A. Bob's task is to guess (with probability of being correct bigger than 1/2) whether $A = a_1$ or $A = a_2$.

Bob's strategy is as follows: after seeing A, Bob draws a number B uniformly at random from $(0, 1)$; if B is smaller than A, Bob declares that $A = a_2$; otherwise, he declares that $A = a_1$.

Denote by E the event that Bob is correct using this strategy.

Then,

$$\begin{aligned}P(E) &= P(E|A=a_1)\cdot P(A=a_1)\\ &\quad + P(E|A=a_2)\cdot P(A=a_2)\\ &= P(B>a_1)\cdot P(A=a_1)\\ &\quad + P(B<a_2)\cdot P(A=a_2)\\ &= (1-a_1)\cdot 0.5 + a_2\cdot 0.5\\ &= 0.5+0.5(a_2-a_1)\\ &> 0.5.\end{aligned}$$

The probability that Bob is correct using this strategy is therefore greater than $\frac{1}{2}$. □

Question 16. How many digits does the number 125^{100} have? You are not allowed to use values of $\log_{10} 2$ or $\log_{10} 5$.

Answer: Note that

$$125^{100} = \left(\frac{1000}{8}\right)^{100} = \frac{1000^{100}}{2^{300}}. \tag{6.199}$$

Since $2^{10} = 1024$, we obtain that

$$125^{100} = \frac{1000^{100}}{1024^{30}} = \frac{1000^{70}}{1.024^{30}}. \tag{6.200}$$

We first show that

$$1 < 1.024^{30} < 10. \tag{6.201}$$

From the binomial expansion, it follows that

$$(1+0.024)^{30} = \sum_{j=0}^{30}\binom{30}{j}0.024^j. \tag{6.202}$$

Note that the ratio of every two consecutive terms in (6.202) is less than 0.72, since

$$\begin{aligned}
\frac{\binom{30}{j+1}0.024^{j+1}}{\binom{30}{j}0.024^j} &= \frac{\binom{30}{j+1}}{\binom{30}{j}} \cdot 0.024 \\
&= \frac{30!}{(j+1)!\,(30-j-1)!} \cdot \frac{j!\,(30-j)!}{30!} \cdot 0.024 \\
&= \frac{30-j}{j+1} \cdot 0.024 \\
&< 30 \cdot 0.024 \\
&= 0.72.
\end{aligned}$$

Then,

$$\binom{30}{j}0.024^j \le 0.72^j \;\; \forall\, 0 \le j \le 30, \qquad (6.203)$$

and, from (6.202) and (6.203), we obtain that

$$\begin{aligned}
(1+0.024)^{30} &< \sum_{j=0}^{30} 0.72^j < \sum_{j=0}^{\infty} 0.72^j \\
&= \frac{1}{1-0.72} \\
&< 10;
\end{aligned}$$

note that, for the equality on the second line above, we used the geometric series identity

$$\sum_{j=1}^{\infty} z^j = \frac{1}{1-z} \quad \text{for } z = 0.72.$$

Thus, the inequality (6.201) is proved.

From (6.200) and (6.201), we obtain that

$$10^{209} < 125^{100} = \frac{10^{210}}{1.024^{30}} < 10^{210}, \qquad (6.204)$$

and conclude that 125^{100} has 210 digits. □

Question 17. For every subset of $\{1, 2, 3, \ldots, 2013\}$, arrange the numbers in the increasing order and take the sum with alternating signs. The resulting integer is called the weight of the subset.[14] Find the sum of the weights of all the subsets of $\{1, 2, 3, \ldots, 2013\}$.

Answer: Let $w(S)$ denote the weight of subset S. Every subset S of $\{1, 2, 3, \ldots, 2013\}$ that does not contain element 1 can be uniquely paired with the subset $\{1\} \cup S$ that contains element 1. Since there are 2^{2013} subsets of $\{1, 2, 3, \ldots, 2013\}$, there are 2^{2012} such pairs. Note that $w(S) + w(\{1\} \cup S) = 1$; that is, the combined weight of each pair is 1. For example,

$$\begin{aligned} & w(\{2,5,8\}) + w(\{1,2,5,8\}) \\ = & (2 - 5 + 8) + (1 - 2 + 5 - 8) \\ = & 1. \end{aligned}$$

Hence, the sum of the weights of all the subsets of $\{1, 2, 3, \ldots, 2013\}$ is 2^{2012}. □

Question 18. Alice and Bob alternately choose one number from one of the following nine numbers: 1/16, 1/8, 1/4, 1/2, 1, 2, 4, 8, 16, without replacement. Whoever gets three numbers that multiply to one wins the game. Alice starts first. What should her strategy be? Can she always win?

Answer: First, notice that the numbers Alice and Bob play with are powers of two, namely, 2^{-4}, 2^{-3}, 2^{-2}, 2^{-1}, 2^0, 2^1, 2^2, 2^3, and 2^4. Next, imagine that Alice and Bob are playing on the 3×3 square, whose entries are as follows: the first row (from left to right) is 2^3, 2^{-4}, 2^1; the

[14]For example, the weight of the subset $\{3\}$ is 3. The weight of the subset $\{2, 5, 8\}$ is $2 - 5 + 8 = 5$.

second row is 2^{-2}, 2^0, 2^2; and the third row is 2^{-1}, 2^4, 2^{-3}.

The product of the entries in every row, every column, and every diagonal is 1, and these possibilities cover all the ways to choose three of the given numbers multiplying to 1. Thus, Alice and Bob are essentially playing Tic-Tac-Toe! It is well-known that best play by both players in Tic-Tac-Toe leads to a draw. We conclude that Alice does not have a winning strategy, although she cannot lose either. □

Question 19. Mr. and Mrs. Jones invite four other couples over for a party. At the end of the party, Mr. Jones asks everyone else how many people they shook hands with, and finds that everyone gives a different answer. Of course, no one shook hands with his or her spouse and no one shook the same person's hand twice. How many people did Mrs. Jones shake hands with?

Answer: Since each person shook hands with at most eight others, the nine different answers received by Mr. Jones are exactly the numbers 0 through 8. Denote by P_i the person with i handshakes, $i = 0, 1, \ldots, 8$. Mr. Jones is not assigned any additional notation.

P_8 shook hands with 8 people of the total of 9 other people. Thus, P_8 did not shake the hand of only one other person, so that person must be his or her spouse. On the other hand, P_8 did not shake the hand of P_0 since nobody did that. Therefore, P_8 and P_0 are married, and P_8 shook everyone's hand except for P_0. P_7 did not shake the hands of two people, one of whom was his/her spouse. One of these two people had to be P_0 as he or she did not shake anyone's hand, and the other one had to be P_1 as he or she had only one handshake, namely with P_8. Since spouses do not shake hands, the spouse of P_7 is either P_1 or P_0. However, P_0 is married to P_8, so P_1 must be married to P_7.

Proceeding similarly, we find that P_6 and P_2 must be married, and that P_5 and P_3 must be married. Then, P_4 must be Mrs. Jones, since this is the only person whose spouse was not identified, and Mr. Jones was not any one of $P_0, \ldots, P_8$.

We conclude that Mrs. Jones shook hands with four people. □

Question 20. The New York Yankees and the San Francisco Giants are playing in the World Series (best of seven format). You would like to bet \$100 on the Yankees winning the World Series, but you can only place bets on individual games, and every time at even odds. How much should you bet on the first game?

Answer: Let P be a 5×5 matrix containing the net payoffs for all the states in this dynamic programming problem. More precisely, $P(i,j)$ denotes the net payoff in the state (i,j) when Yankees have won i and lost j games ($0 \le i,j \le 4$). Clearly, $P(4,j) = 100$ ($0 \le j \le 3$), and $P(i,4) = -100$ ($0 \le i \le 3$). Moreover, $P(4,4)$ is left blank, as $(4,4)$ is not in our state space (4 wins and 4 losses cannot be achieved in a best of seven series).

Let B be a 4×4 matrix containing the bets we need to place at each state (i,j), given that we would like to bet \$100 on the Yankees winning the World Series, and given that Yankees have won i and lost j games so far ($0 \le i,j \le 3$). Clearly, $B(3,3) = 100$.

Given that Yankees have won i and lost j games so far, if we bet $B(i,j)$ on the Yankees for the next game, our payoff will be $P(i,j) + B(i,j)$ if the Yankees win, or $P(i,j) - B(i,j)$ if the Yankees lose. Therefore,

$$P(i+1,j) = P(i,j) + B(i,j); \qquad (6.205)$$

$$P(i,j+1) = P(i,j) - B(i,j). \qquad (6.206)$$

By adding and subtracting the equations (6.205) and

(6.206) we obtain that

$$P(i,j) = \frac{1}{2}(P(i+1,j) + P(i,j+1)) \quad (6.207)$$

$$B(i,j) = \frac{1}{2}(P(i+1,j) - P(i,j+1)) \quad (6.208)$$

Now, it is easy to compute all the entries of P, using (6.207) and working backwards from $P(4,j)$ and $P(i,4)$. For example,

$$P(3,3) = \frac{1}{2}(P(4,3) + P(3,4)) = \frac{1}{2}(100 - 100) = 0.$$

Once matrix P is computed, we use (6.208) to compute the matrix B. For example,

$$B(2,1) = \frac{1}{2}(P(3,1) - P(2,2)) = \frac{1}{2}(75 - 0) = 37.5.$$

In other words, given that Yankees have won 2 games and lost 1 game so far, and we would like to bet \$100 on the Yankees winning the World Series, we should bet \$37.5 on the Yankees for the next game. We include both matrices below:

$$P = \begin{pmatrix} 0 & -31.25 & -62.5 & -87.5 & -100 \\ 31.25 & 0 & -37.5 & -75 & -100 \\ 62.5 & 37.5 & 0 & -50 & -100 \\ 87.5 & 75 & 50 & 0 & -100 \\ 100 & 100 & 100 & 100 & 0 \end{pmatrix};$$

$$B = \begin{pmatrix} 31.25 & 31.25 & 25 & 12.5 \\ 31.25 & 37.5 & 37.5 & 25 \\ 25 & 37.5 & 50 & 50 \\ 12.5 & 25 & 50 & 100 \end{pmatrix}.$$

Therefore, we should bet $B(1,1) = 31.25$ dollars (on the Yankees) on the first game. Note that the probability of the Yankees winning or losing a single game does not affect your betting strategy. □

Question 21. We have two red, two green and two yellow balls. For each color, one ball is heavy and the other is light. All heavy balls weigh the same. All light balls weigh the same. How many weighings on a scale are necessary to identify the three heavy balls?

Answer: It is clear that one weighing does not suffice. We show that the heavy balls can be identified in only two weighings.

Label the red balls R_1 and R_2, the green balls G_1 and G_2, and the yellow balls Y_1 and Y_2. Our first weighing is $\{R_1, G_1\}$ vs. $\{R_2, Y_1\}$.

If the scale is in balance, then either G_1 or Y_1 is heavy, but not both. Our second weighing is $\{G_1\}$ vs. $\{Y_1\}$. If G_1 is heavier, then the set of heavy balls is $\{R_2, G_1, Y_2\}$. If Y_1 is heavier, then the set of heavy balls is $\{R_1, G_2, Y_1\}$.

If $\{R_1, G_1\}$ is heavy, then either G_1 is heavy or Y_1 is light. Our second weighing is $\{G_1, Y_1\}$ vs. $\{G_2, Y_2\}$. If the scale is in balance, then G_1 is heavy; hence, the set of heavy balls is $\{R_1, G_1, Y_2\}$. If $\{G_1, Y_1\}$ is heavier, then G_1 and Y_1 are both heavy. The set of heavy balls is $\{R_1, G_1, Y_1\}$. If $\{G_2, Y_2\}$ is heavier, then G_2 and Y_2 are both heavy. The set of heavy balls is $\{R_1, G_2, Y_2\}$.

If $\{R_2, Y_1\}$ is heavy, then either Y_1 is heavy or G_1 is light. Our second weighing is $\{G_1, Y_1\}$ vs. $\{G_2, Y_2\}$. If the scale is in balance, then Y_1 is heavy; hence, the set of heavy balls is $\{R_2, G_2, Y_1\}$. If $\{G_1, Y_1\}$ is heavy, then G_1 and Y_1 are both heavy. The set of heavy balls is $\{R_2, G_1, Y_1\}$. If $\{G_2, Y_2\}$ is heavier, then G_2 and Y_2 are both heavy. The set of heavy balls is $\{R_2, G_2, Y_2\}$. □

Question 22. There is a row of 10 rooms and a treasure in one of them. Each night, a ghost moves the treasure to an adjacent room. You are trying to find the treasure, but can only check one room per day. How do you find it?

Answer: The treasure can be found in at most 16 days. Label the rooms 1 through 10. Denote by T_k the room where the treasure is on day k, and let R_k denote the room you check on day k. Adopt the following strategy: for days $k = 1, 2, \ldots, 8$, let $R_k = k + 1$; for days $k = 9, 10, \ldots, 16$, let $R_k = 18 - k$.

If T_1 is even, then we will find the treasure in one of the first eight days. In other words, there exists k with $1 \leq k \leq 8$ such that $T_k = R_k$. Note that for $1 \leq k \leq 8$, T_k and R_k have the same parity, since T_1 is even, $R_1 = 2$, and both T_k and R_k change by at most 1 from day to day, according to ghost's moves and our strategy. Hence, $T_k - R_k$ is even. Furthermore, $T_1 \neq 1$ and $T_8 \neq 10$, which implies $T_1 - R_1 \geq 0$ and $T_8 - R_8 \leq 0$. Since $T_k - R_k$ can change by at most 2 from day to day, there must exist some k, $1 \leq k \leq 8$, such that $T_k - R_k = 0$.

If T_1 is odd, then we claim $T_k = R_k$ for some k, $9 \leq k \leq 16$. Note that T_9 is odd since T_1 is odd and the treasure is moved to an adjacent room each night. Furthermore, for $9 \leq k \leq 16$, T_k and R_k have the same parity since T_9 is odd, $R_9 = 9$, and both T_k and R_k change by at most 1 from day to day, according to ghost's moves and our strategy. Hence, $R_k - T_k$ is even. Furthermore, $T_9 \neq 10$ and $T_{16} \neq 1$, which implies $R_9 - T_9 \geq 0$ and $R_{16} - T_{16} \leq 0$. Since $R_k - T_k$ can change by at most 2 from day to day, there must exist some k, $9 \leq k \leq 16$, such that $R_k - T_k = 0$.

Finally, note that this strategy can be generalized to any number $n \geq 2$ of rooms, when the treasure will be found in $2n - 4$ days, by checking the rooms $2, 3, \ldots, n - 1, n - 1, n - 2, \ldots, 3, 2$, in that order. □

Question 23. How many comparisons do you need to find the maximum in a set of n distinct numbers? How many comparisons do you need to find both the maximum and minimum in a set of n distinct numbers?

Answer: We can find the maximum in $n-1$ comparisons as follows: let $\{x_1, x_2, \ldots, x_n\}$ denote the set of n distinct numbers. Scan the numbers from left to right, while maintaining the current maximum M. More precisely, set $x_1 = M$, and, in the ith comparison, compare M and x_{i+1}. If $M > x_{i+1}$, leave M as is; otherwise, set $M = x_{i+1}$. Do this for $i = 1, \ldots, n-1$.

Similarly, one can find the minimum in a set of n distinct numbers in $n-1$ comparisons.

Then, we can find the maximum M and the minimum m in $\{x_1, x_2, \ldots, x_n\}$ with $2n-3$ comparisons: find the maximum M in $\{x_1, x_2, \ldots, x_n\}$ with $n-1$ comparisons, and then find the minimum m in $\{x_1, x_2, \ldots, x_n\} \setminus \{M\}$ with $n-2$ comparisons.

However, one can find m and M significantly faster.

If n is even, compare all $\frac{n}{2}$ consecutive pairs of numbers x_{2i-1} and x_{2i}, $i = 1 : \frac{n}{2}$, and put the smaller number into a set S and the larger number into a set L. This requires $\frac{n}{2}$ comparisons. Note that S and L have $\frac{n}{2}$ elements each. Then, find the minimum m in S using $\frac{n}{2} - 1$ comparisons, and find the maximum M in L using $\frac{n}{2} - 1$ comparisons. The total number of comparisons to find m and M is

$$\frac{n}{2} + \left(\frac{n}{2} - 1\right) + \left(\frac{n}{2} - 1\right) = \frac{3n}{2} - 2. \qquad (6.209)$$

If n is odd, compare $\frac{n-1}{2}$ consecutive pairs of numbers x_{2i-1} and x_{2i}, $i = 1 : \frac{n-1}{2}$, and put the smaller number into a set S and the larger number into a set L. This requires $\frac{n-1}{2}$ comparisons. Place x_n into both S and L. Note that S and L have $\frac{n-1}{2} + 1 = \frac{n+1}{2}$ elements each. Then, find the minimum m in S using $\frac{n+1}{2} - 1 = \frac{n-1}{2}$ comparisons, and find the maximum M in L using $\frac{n+1}{2} - 1 = \frac{n-1}{2}$ comparisons. The total number of comparisons to find m and M is

$$\frac{n-1}{2} + \frac{n-1}{2} + \frac{n-1}{2} = \frac{3n+1}{2} - 2. \qquad (6.210)$$

Note that the results of (6.209) and (6.210) can be written succinctly as

$$\left\lceil \frac{3n}{2} \right\rceil - 2 \quad \text{comparisons},$$

where $\lceil x \rceil$ denotes the ceiling of x, the smallest integer greater than or equal to x. □

Question 24. Given a cube, you can jump from one vertex to a neighboring vertex with equal probability. Assume you start from a certain vertex (does not matter which one). What is the expected number of jumps to reach the opposite vertex?

Answer: Label the vertices of the cube with 0, 1, 2, and 3, according to your distance from the opposite vertex. In other words, label the starting vertex with 3, the vertices adjacent to the starting vertex with 2, the vertices adjacent to the opposite vertex with 1, and the opposite vertex with 0. Call the opposite vertex your final destination.

Denote by E_i, $i = 0, 1, 2, 3$, the expected number of jumps yet to be made to reach the final destination, given that you are currently in one of the vertices labeled with i. Note that $E_0 = 0$, and we have to find E_3.

After the first jump, you are in one of the vertices labeled with 2, so

$$E_3 = 1 + E_2. \qquad (6.211)$$

From a vertex labeled with 2, you can jump to three vertices: two of them are labeled with 1, and one of them is labeled with 3. Thus, you jump to a vertex labeled with

1 with probability $\frac{2}{3}$, or to a vertex labeled with 3 with probability $\frac{1}{3}$. Hence,

$$E_2 = 1 + \frac{2}{3}E_1 + \frac{1}{3}E_3. \qquad (6.212)$$

Similarly, from one of the vertices labeled with 1, you jump to a vertex labeled with 2 with probability $\frac{2}{3}$, or to a vertex labeled with 0 with probability $\frac{1}{3}$. Hence,

$$E_1 = 1 + \frac{2}{3}E_2 + \frac{1}{3}E_0 = 1 + \frac{2}{3}E_2, \qquad (6.213)$$

since $E_0 = 0$.

Solving (6.211–6.213) yields $E_1 = 7$, $E_2 = 9$, and $E_3 = 10$.

We conclude that it will take 10 jumps, on average, to reach the opposite vertex. □

Question 25. Select numbers uniformly distributed between 0 and 1, one after the other, as long as they keep decreasing; i.e. stop selecting when you obtain a number that is greater than the previous one you selected.

(i) On average, how many numbers have you selected?

(ii) What is the average value of the smallest number you have selected?

Answer: We give three solutions for part (i); the third solution will be used to solve part (ii).

(i) *Solution 1:* Denote by $E(x)$ the expected number of numbers you have yet to select, given that you have just selected number x. For example, $E(0) = 1$, since the next number you select is greater than $x = 0$, upon which the game stops.

Assume that you have just selected number x. Denote by y the next number you select. We find $E(x)$ by conditioning on y. With probability $1 - x$, y is greater than x; the game stops, and, thus, you have selected only one number after selecting x. On the other hand, y could be smaller than x, in which case you expect to select $E(y)$ additional numbers after selecting y; in other words, you expect to select $1+E(y)$ additional numbers after selecting x. Since the probability density function of y is $f(y) = 1$, the law of total probability gives

$$\begin{aligned} E(x) &= 1 \cdot (1 - x) + \int_0^x (1 + E(y)) f(y)\, dy \\ &= (1 - x) + \int_0^x (1 + E(y))\, dy \\ &= 1 + \int_0^x E(y)\, dy. \end{aligned}$$

Differentiating

$$E(x) = 1 + \int_0^x E(y)\, dy$$

with respect to x yields

$$E'(x) = E(x).$$

Thus, $E(x) = Ce^x$, where C is a constant. The condition $E(0) = 1$ gives $C = 1$. Hence, $E(x) = e^x$.

Note that the first number selected is automatically smaller than 1. Then, the number of numbers selected after starting with the number 1, which was denoted by $E(1)$, is equal to the number of numbers selected starting with a random number between 0 and 1. Therefore, the average number of numbers you have selected is $E(1) = e$.

(i) *Solution 2:* Denote by N the average number of numbers you have selected. Denote by x_i the ith number you selected, $i \geq 1$, and let p_i denote the probability that $x_i < x_{i-1} < \ldots < x_1$. Since there are $i!$ permutations of $x_1, \ldots, x_i$, then $p_i = \frac{1}{i!}$.

You select at least two numbers before stopping. The probability that you select exactly i numbers, $i \geq 2$, before stopping is equal to the probability that $x_{i-1} < \ldots < x_2 < x_1$ and $x_i > x_{i-1}$. The latter equals the probability that $x_{i-1} < \ldots < x_2 < x_1$ minus the probability that $x_i < x_{i-1} < \ldots < x_2 < x_1$; that is, $p_{i-1} - p_i$.

We conclude that the expected number of numbers you have selected is

$$\begin{aligned} N &= \sum_{i=2}^{\infty} i(p_{i-1} - p_i) \\ &= \sum_{i=2}^{\infty} i\left(\frac{1}{(i-1)!} - \frac{1}{i!}\right) \\ &= \sum_{i=2}^{\infty} i \cdot \frac{i-1}{i!} \\ &= \sum_{i=2}^{\infty} \frac{1}{(i-2)!} \\ &= \sum_{j=0}^{\infty} \frac{1}{j!} \\ &= e, \end{aligned} \tag{6.214}$$

where (6.214) follows from the Taylor series expansion of e^x around 0, i.e.

$$e^x = \sum_{j=0}^{\infty} \frac{x^j}{j!}, \quad \forall\, x \in \mathbb{R},$$

by letting $x = 0$.

(i) *Solution 3:* Denote by $p(x)\,dx$ the probability that a number between x and $x + dx$ is selected as part of the decreasing sequence. Let $p_i(x)\,dx$ denote the probability that a number between x and $x + dx$ is selected as the ith term of the decreasing sequence. Then,

$$p(x)\,dx = \left(\sum_{i=1}^{\infty} p_i(x)\right)\,dx. \tag{6.215}$$

The probability that a number between x and $x + dx$ is selected as the ith term of the decreasing sequence is

$$p_i(x)\,dx = \frac{1}{(i-1)!}(1-x)^{i-1}\,dx, \tag{6.216}$$

since $(1-x)^{i-1}$ is the probability that the first $i-1$ numbers selected are greater than x, and $\frac{1}{(i-1)!}$ is the probability that they are selected in decreasing order.

From (6.215) and (6.216), we obtain that

$$\begin{aligned} p(x)\,dx &= \sum_{i=1}^{\infty} \frac{(1-x)^{i-1}}{(i-1)!}\,dx \\ &= \sum_{j=0}^{\infty} \frac{(1-x)^j}{j!}\,dx \\ &= e^{1-x}\,dx; \end{aligned} \tag{6.217}$$

where (6.217) follows from the Taylor series expansion of e^t around 0, i.e.

$$e^t = \sum_{j=0}^{\infty} \frac{t^j}{j!}, \quad \forall\, t \in \mathbb{R},$$

by letting $t = 1 - x$.

Therefore, the expected number of numbers selected in the decreasing sequence is

$$\int_0^1 p(x)\,dx = \int_0^1 e^{1-x}\,dx = e - 1.$$

Adding the last number selected (which is not in the decreasing sequence) gives an average of e numbers selected.

(ii) Denote by s the smallest number selected. It is the last number selected in the decreasing sequence. Since the probability that a number between x and $x + dx$ is selected as part of the decreasing sequence equals $e^{1-x}\,dx$, see (6.217), and since the probability that the next number selected is larger is $(1-x)$, then the probability that s is between x and $x+dx$ is $e^{1-x}(1-x)\,dx$.

Therefore, the expected value of the smallest number you have selected is

$$\begin{aligned} E(s) &= \int_0^1 xe^{1-x}(1-x)\,dx \\ &= \int_0^1 e^y y(1-y)\,dy, \qquad (6.218) \end{aligned}$$

where (6.218) follows from the substitution $y=1-x$.

Using integration by parts to compute (6.218), we find that

$$\begin{aligned} E(s) &= e^y(y-y^2)\Big|_0^1 - \int_0^1 e^y(1-2y)\,dy \\ &= e^y(y-y^2)\Big|_0^1 - e^y\Big|_0^1 + 2\int_0^1 ye^y\,dy \\ &= e^y(y-y^2-1)\Big|_0^1 + 2ye^y\Big|_0^1 - 2\int_0^1 e^y\,dy \\ &= e^y(3y-y^2-1)\Big|_0^1 - 2e^y\Big|_0^1 \\ &= e^y(3y-y^2-3)\Big|_0^1 \\ &= 3-e. \quad \square \end{aligned}$$

Question 26. To organize a charity event that costs \$100K, an organization raises funds. Independent of each other, one donor after another donates some amount of money that is exponentially distributed with a mean of \$20K. The process is stopped as soon as \$100K or more has been collected. Find the distribution, mean, and variance of the number of donors needed until at least \$100K has been collected.

Answer: Denote by a_i the amount of money donated by donor i, that is exponentially distributed with mean $1/\lambda$. Let $s_n = \sum_{i=1}^{n} a_i$ be the total amount of money donated by donors $1, \ldots, n$, and let

$$N = \min_{n \geq 1} \{ n \text{ such that } s_n \geq a \}$$

be the discrete random variable denoting the smallest index n such that s_n is at least a.

Denote by $P(n|a)$, $n \geq 1$, the probability mass function of N, that is, the probability that $N = n$ when a total of a needs to be raised. Note that

$$P(1|a) = P(a_1 \geq a) = e^{-\lambda a}. \qquad (6.219)$$

We find $P(n|a)$, $n > 1$, by conditioning on a_1. Given that the first donor donated $a_1 = x < a$, N is equal to n if and only if the remaining amount $a - x$ is raised by the next $n - 1$ donors (and not by fewer than the next $n - 1$ donors), an event that by definition has probability $P(n-1|a-x)$. Since the probability density function of a_1 is $f_{a_1}(x) = \lambda e^{-\lambda x}$, then, for $n > 1$, the law of total probability yields

$$\begin{aligned} P(n|a) &= \int_0^a P(n-1|a-x) f_{a_1}(x)\, dx \\ &= \int_0^a \lambda e^{-\lambda x} P(n-1|a-x)\, dx. \qquad (6.220) \end{aligned}$$

We will prove that

$$P(n|a) = \frac{(\lambda a)^{n-1}}{(n-1)!}\, e^{-\lambda a}, \;\; \forall\, a \geq 0, \qquad (6.221)$$

by induction on n. The base case $n = 1$ was already established; see (6.219). Assume that (6.221) holds for $n > 1$; we will show that it also holds for $n + 1$.

From the induction hypothesis, we obtain that

$$P(n|a-x) = \frac{(\lambda(a-x))^{n-1}}{(n-1)!} e^{-\lambda(a-x)}, \qquad (6.222)$$

for all $0 \leq x \leq a$. From (6.220), it follows that

$$P(n+1|a) = \int_0^a \lambda e^{-\lambda x} P(n|a-x)\, dx. \qquad (6.223)$$

From (6.222) and (6.223), we find that

$$\begin{aligned} & P(n+1|a) \\ = & \int_0^a \lambda e^{-\lambda x} \cdot \frac{(\lambda(a-x))^{n-1}}{(n-1)!} e^{-\lambda(a-x)}\, dx \\ = & \frac{\lambda^n e^{-\lambda a}}{(n-1)!} \int_0^a (a-x)^{n-1}\, dx \\ = & \frac{(\lambda a)^n}{n!} e^{-\lambda a}. \end{aligned}$$

We conclude that (6.221) holds for $n+1$, and therefore (6.221) is proved by induction.

From (6.221), it follows that N has the same distribution as $1+M$, where M has a Poisson distribution with mean λa. Then,

$$E[N] = 1 + \lambda a; \quad \mathrm{Var}(N) = \lambda a.$$

For our problem, $1/\lambda = \$20\mathrm{K}$ and $a = \$100\mathrm{K}$. Thus, $E[N] = 1 + \lambda a = 6$ and $\mathrm{Var}(N) = \lambda a = 5$.

We conclude that the number of donors needed until at least $100K is collected has mean 6 and variance 5. □

Question 27. Consider a random walk starting at 1 and with equal probability of moving to the left or to the right by one unit, and stopping either at 0 or at 3.

(i) What is the expected number of steps to do so?

(ii) What is the probability of the random walk ending at 3 rather than at 0?

Answer: Denote by X_n^ℓ the position of the random walk at time n, where the superscript refers to the starting position of the walk; for example, $X_0^\ell = \ell$. In this question, we are concerned with X_n^1. For a random walk starting at $\ell \in \{0, 1, 2, 3\}$, denote by T_ℓ the number of steps taken by the random walk in order to reach either 0 or 3 for the first time, and by $t_\ell = E[T_\ell]$ the expected number of steps.

(i) We are looking to find t_1. We derive a recurrence relation among the t_ℓ's as follows: for $\ell = 1, 2$,

$$\begin{aligned} t_\ell &= E[T_\ell] \\ &= E[T_\ell | X_1^\ell = \ell - 1] \cdot P(X_1^\ell = \ell - 1) \\ &\quad + E[T_\ell | X_1^\ell = \ell + 1] \cdot P(X_1^\ell = \ell + 1) \\ &= E[1 + T_{\ell-1}] \cdot P(X_1^\ell = \ell - 1) \qquad (6.224) \\ &\quad + E[1 + T_{\ell+1}] \cdot P(X_1^\ell = \ell + 1) \qquad (6.225) \\ &= (1 + t_{\ell-1}) \cdot \frac{1}{2} + (1 + t_{\ell+1}) \cdot \frac{1}{2}, \qquad (6.226) \end{aligned}$$

where the following identities were used to derive (6.224) and (6.225):

$$E[T_\ell | X_1^\ell = \ell - 1] = E[1 + T_{\ell-1}]; \qquad (6.227)$$

$$E[T_\ell | X_1^\ell = \ell + 1] = E[1 + T_{\ell+1}]. \qquad (6.228)$$

Note that (6.227) and (6.228) follow from the fact that, starting at ℓ, once the random walk took its first step to $\ell - 1$ (or $\ell + 1$), it becomes equivalent to a random walk starting afresh at $\ell - 1$ (or $\ell + 1$). The plus 1 term on the right hand side of (6.227) and (6.228) accounts for the first step.

Since $t_0 = t_3 = 0$, by letting $l = 1$ and then $l = 2$ in (6.226), we obtain the following linear system for t_1 and

t_2:

$$\begin{aligned} t_1 &= \frac{1+t_0}{2} + \frac{1+t_2}{2} \\ &= \frac{1}{2} + \frac{1+t_2}{2}; \end{aligned}$$

$$\begin{aligned} t_2 &= \frac{1+t_1}{2} + \frac{1+t_3}{2} \\ &= \frac{1+t_1}{2} + \frac{1}{2}. \end{aligned}$$

Thus, $t_1 = 2$ and $t_2 = 2$.

We conclude that the expected number of steps before stopping is 2.

(ii) Denote by p_ℓ the probability that the random walk reaches 3 before it reaches 0 when its position is at ℓ, for $\ell \in \{0, 1, 2, 3\}$. We need to find p_1. Denote by τ_0^ℓ the first time the walk reaches 0 when starting at ℓ, and by τ_3^ℓ the first time the walk reaches 3 when starting at ℓ. We derive a recurrence relation for the p_ℓ's as follows: for $\ell = 1, 2$,

$$\begin{aligned} p_\ell &= P\left(\tau_3^\ell < \tau_0^\ell\right) \\ &= P\left(\tau_3^\ell < \tau_0^\ell \middle| X_1^\ell = \ell+1\right) P(X_1^\ell = \ell+1) \\ &\quad + P\left(\tau_3^\ell < \tau_0^\ell \middle| X_1^\ell = \ell-1\right) P(X_1^\ell = \ell-1) \\ &= P\left(\tau_3^{\ell+1} < \tau_0^{\ell+1}\right) P(X_1^\ell = \ell+1) \\ &\quad + P\left(\tau_3^{\ell-1} < \tau_0^{\ell-1}\right) P(X_1^\ell = \ell-1) \\ &= \frac{1}{2} \cdot p_{\ell+1} + \frac{1}{2} \cdot p_{\ell-1}, \end{aligned} \tag{6.229}$$

because the random walk is equally likely to move to the left or to the right and, once it moved, the random walk starts afresh.

Since $p_0 = 0$ and $p_3 = 1$, by letting $l = 1$ and then $l = 2$ in (6.229), we obtain the following linear system for p_1 and p_2:

$$\begin{aligned} p_1 &= \frac{p_0}{2} + \frac{p_2}{2} = \frac{p_2}{2}; \\ p_2 &= \frac{p_1}{2} + \frac{p_3}{2} = \frac{p_1}{2} + \frac{1}{2}. \end{aligned}$$

Thus, $p_1 = \frac{1}{3}$ and $p_2 = \frac{2}{3}$.

We conclude that the probability of the random walk ending at 3 rather than at 0 is $\frac{1}{3}$. □

Question 28. A stick of length 1 drops and breaks at a random place uniformly distributed across the length. What is the expected length of the smaller part?

Answer:

Solution 1: Treating the stick as an interval $[0, 1]$, the breakpoint X becomes a random variable uniformly distributed on $(0, 1)$. Its probability density function $f_X(x)$ is 1, for $0 \leq x \leq 1$, and 0 otherwise. Denote by L the length of the smaller part. Then, $L = \min\,(x, 1-x)$. We conclude that

$$\begin{aligned} E[L] &= \int_0^1 \min\,(x, 1-x) \cdot f_X(x)\,dx \\ &= \int_0^1 \min\,(x, 1-x)\,dx \\ &= \int_0^{1/2} x\,dx + \int_{1/2}^1 (1-x)\,dx \\ &= \frac{1}{8} + \frac{1}{8} \\ &= \frac{1}{4}. \end{aligned}$$

Solution 2: Treating the stick as an interval $[0, 1]$, the breakpoint X becomes a random variable uniformly distributed on $(0,1)$. Denote by L the length of the smaller part.

Let A be the event that the breakpoint X is in $\left(0, \frac{1}{2}\right)$. Then, $\overline{A}$ is the event that the breakpoint X is in $\left(\frac{1}{2}, 1\right)$. Clearly, $P(A) = P(\overline{A}) = \frac{1}{2}$. Given A, the length of the smaller part is X. Given $\overline{A}$, the length of the smaller part is $1 - X$.

The law of total probability yields

$$\begin{aligned} E[L] &= E[L|A] \cdot P(A) + E[L|\overline{A}] \cdot P(\overline{A}) \\ &= E[X|A] \cdot \frac{1}{2} + E[1 - X|\overline{A}] \cdot \frac{1}{2} \\ &= E[X|A] \cdot \frac{1}{2} + \frac{1}{2} - E[X|\overline{A}] \cdot \frac{1}{2}. \quad (6.230) \end{aligned}$$

Note that, given A, X is uniformly distributed on $\left(0, \frac{1}{2}\right)$, and, Given $\overline{A}$, X is uniformly distributed on $\left(\frac{1}{2}, 1\right)$. Since the expectation of a random variable uniformly distributed on an interval (a, b) is equal to $\frac{a+b}{2}$, we obtain that

$$E[X|A] = \frac{1}{4} \quad \text{and} \quad E[X|\overline{A}] = \frac{3}{4}.$$

Then, (6.230) yields

$$\begin{aligned} E[L] &= \frac{1}{4} \cdot \frac{1}{2} + \frac{1}{2} - \frac{3}{4} \cdot \frac{1}{2} \\ &= \frac{1}{4}. \quad \square \end{aligned}$$

Question 29. You are given a stick of unit length.

(i) The stick drops and breaks at two places. What is the probability that the three pieces could form a triangle?

(ii) The stick drops and breaks at one place. Then the larger piece is taken and dropped again, breaking at one place. What is the probability that the three pieces could form a triangle?

Answer: We offer two solutions for part (i); the second solution will be used to solve part (ii).

(i) *Solution 1:* Denote by X and Y the two break points, and assume X and Y are independent random variables uniformly distributed on $(0, 1)$. To form a triangle, the sum of the lengths of any two pieces must be greater than the length of the third piece. Equivalently, each piece must be of length less than $1/2$.

Assume that $Y > X$. Then, the length of the three pieces are X, $Y - X$, and $1 - Y$. Each of these pieces is of length less than $1/2$ if and only if the point (X, Y) belongs to the region $\{(x, y) : x < 1/2, y - x < 1/2, 1 - y < 1/2, x \in (0, 1), y \in (0, 1), x < y\}$ in the unit square. Since the area of this region is $1/8$, the probability of the three pieces forming a triangle, given $Y > X$, is $1/8$. Symmetrically, the probability of the three pieces forming a triangle, given $Y < X$, is $1/8$. Events $\{Y > X\}$ and $\{Y < X\}$ are disjoint; hence, the probability of the three pieces forming a triangle is $1/4$.

(i) *Solution 2:* Consider an equilateral triangle ABC with height of length 1. Given a point P in its interior, let h_a, h_b, and h_c be the lengths of the perpendiculars dropped from P to the sides BC, CA, and AB, respectively. Since the areas of triangles BPC, CPA, and APB sum up to the area of ABC, we conclude that $h_a + h_b + h_c = 1$, and, thus, is independent of the position of P. Breaking a stick of length 1 into three pieces of lengths h_a, h_b, and h_c, is clearly equivalent to (uniquely) specifying a point P in the interior of the triangle ABC.

Connect the midpoints A', B', C', of the sides of triangle ABC to split it into four congruent equilateral triangles, with the medial triangle $A'B'C'$ in the middle (see Figure 6.3). Each piece of the broken stick has length less than $1/2$ if and only if the corresponding point P belongs to the medial triangle. Since the area of the medial triangle is $1/4$ of the area of triangle ABC, the desired probability is 1/4.

(ii) Assume that the pieces have lengths h and $(1-h)$ after the first break, with $h < (1-h)$, i.e., $h < 1/2$. With h fixed, the (larger) piece of length $1-h$ is taken and dropped again, breaking at one place uniformly at random. The probability that the three pieces thus obtained form a triangle clearly depends on h. It is close to 0 for h close to 0, and it is close to 1 for h close to 1/2.

More precisely, using the representation from the second solution of part (a) above, the probability that the three pieces form a triangle, given a fixed $h < 1/2$, is equal to the probability that the point P, that lies on the segment UZ parallel to side AB and at distance h from it, belongs to the segment VW, that is the intersection of UZ with the medial triangle $A'B'C'$ (see Figure 6.3). Since P is chosen (by the second break of the stick) from UZ uniformly at random, the probability that P belongs to VW is equal to the ratio of their lengths, namely $\frac{VW}{UZ}$.

Next, we express this ratio in terms of h. First, note that the medial triangle $A'B'C'$ has side length $AB/2$ and height 1/2. Since the triangles ABC and UZC are similar, we have $\frac{UZ}{AB} = \frac{1-h}{1}$. Since the triangles WVC' and $A'B'C'$ are similar, we have $\frac{VW}{A'B'} = \frac{h}{1/2}$. Dividing the last two equations, we obtain that

$$\frac{VW}{UZ} = \frac{h}{1-h}.$$

Therefore, given $h < 1/2$, the probability that the three pieces form a triangle is

$$\int_0^{1/2} \frac{h}{1-h}\,dh = -\int_0^{1/2} \frac{1-h}{1-h}\,dh + \int_0^{1/2} \frac{1}{1-h}\,dh.$$

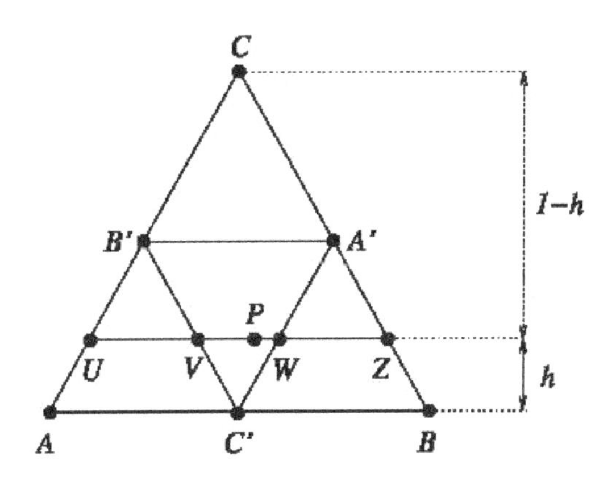

图 6.3: Point P inside the medial triangle $A'B'C'$ of the equilateral triangle ABC.

By using the substitution $y = 1 - h$, it follows that

$$\begin{aligned}\int_0^{1/2} \frac{h}{1-h}\,dh &= -\frac{1}{2} + \int_{1/2}^{1} \frac{1}{y}\,dy \\ &= -\frac{1}{2} - \ln\left(\frac{1}{2}\right) \\ &= \ln 2 - \frac{1}{2}.\end{aligned}$$

Since the probability that $h < 1/2$, where h is chosen uniformly at random from $(0,1)$ (by the first break of the stick), is $1/2$, then the total probability that the three pieces form a triangle is

$$\frac{\ln 2 - 1/2}{1/2} = 2\ln 2 - 1. \quad \square$$

Question 30. Why is a manhole cover round?

Answer: A circle is the shape with minimal surface given a required minimal width in any direction. Moreover, the cover of a round manhole cannot fall through the hole. If the manhole were square, its cover turned on its edge could fall through the hole since the diagonal of a square is $\sqrt{2}$ times larger than its edge. □

Question 31. When is the first time after 12 o'clock that the hour and minute hands of a clock meet again?

Answer: The minute hand moves at a speed of 360 degrees per hour, while the hour hand moves at a speed of 30 degrees per hour. They start together at 12 o'clock. The first time they meet, the minute hand made one full rotation more than the hour hand, which is the same as 360 degrees more than the hour hand. If t denotes the time (measured in hours) until the two hands meet again, this can be written as

$$360 \cdot t \;=\; 30 \cdot t + 360.$$

Thus, $t = \frac{12}{11}$ hours, which is approximately 1 hour, 5 minutes, and 27 seconds. □

Question 32. Three light switches are in one room, and they turn three light bulbs in another. How do you figure out which switch turns on which bulb in one shot?

Answer: Turn on two switches for a couple of minutes, and then turn one of the switches off and go into the other room. The bulb that is lit corresponds to the switch that is still on; the bulb that is not lit but is hot corresponds to the switch that was turned on and then turned off; the bulb that is not lit and is cold corresponds to the switch that was never turned on. □

参考文献

[1] Avner Friedman. *Stochastic Differential Equations and Applications.* Dover Publications, Mineola, New York, 2006.

[2] Paul Glasserman. *Monte Carlo Methods in Financial Engineering.* Springer-Verlag New York, Inc., New York, 2004.

[3] Dan Stefanica. *A Primer For The Mathematics Of Financial Engineering.* Financial Engineering Advanced Background Series. FE Press, New York, 2nd edition, 2011.

[4] Paul Wilmott. *Frequently Asked Questions in Quantitative Finance.* John Wiley & Sons Ltd, Chichester, West Sussex, 2nd edition, 2009.

Made in the USA
Middletown, DE
16 May 2024

54391817R00235